Cognitive Radio

Cognitive Radio
Computing Techniques, Network Security, and Challenges

Edited by
Budati Anil Kumar, Peter Ho Chiung Ching and Shuichi Torii

CRC Press
Taylor & Francis Group
Boca Raton London New York

CRC Press is an imprint of the
Taylor & Francis Group, an **informa** business

First edition published 2022
by CRC Press
6000 Broken Sound Parkway NW, Suite 300, Boca Raton, FL 33487-2742

and by CRC Press
4 Park Square, Milton Park, Abingdon, Oxon, OX14 4RN

ISBN: 9780367609412 (hbk)
ISBN: 9781032147048 (pbk)
ISBN: 9781003102625 (ebk)

DOI: 10.1201/9781003102625

Typeset in Caslon
by codeMantra

Contents

Editors

Budati Anil Kumar is a full-time associate professor at the Gokaraju Rangaraju Institute of Engineering & Technology (GRIET-Autonomous), Hyderabad, India. His research interests mainly focus on wireless communications, signal processing and cognitive radio networks. He has published more than 40 peer-reviewed international journal and conferences papers, delivered guest lectures and acted as various committee chairs in national and international conferences. He acts as guest editor for various highly indexed and high-impact factor journals. He also received funds from AICTE, India to conduct a faculty development program. He is a member of the IEEE, IEI and IAENG since 2018 and has served as a reviewer for various international journals. He has received a B.Tech degree from (VYCET) JNTU, Hyderabad in 2007, a Master of Technology degree from (LBRCE) JNTUK, Kakinada University in 2010 and a Ph.D degree from GITAM Deemed to be University, India in 2019, respectively.

Peter Ho Chiung Ching received his Ph.D. in Information Technology from the Faculty of Computing and Informatics, Multimedia University. His doctoral research work was on the performance evaluation of multimodal biometric systems using fusion techniques. Dr. Ho is a senior member of the Institute of Electrical and Electronics

Engineers. Dr. Ho has published a number of peer-reviewed papers related to location intelligence, multimodal biometrics, action recognition and text mining. He is currently an adjunct senior research fellow in the Department of Computing and Information Systems, School of Science and Technology, Sunway University.

Shuichi Torii received his B.D. degree from Kagoshima University in 1983 and M.D. and Ph.D. degrees from Kyushu University in 1985 and 1989, respectively, all in Mechanical Engineering. He then worked as a visiting scholar at the University of Michigan, where he studied the solidification and oxidization in the reactor using the experimental method and numerical simulation. In 1993, he became the associate professor at Kagoshima University, where he studied the thermal fluid flow transport phenomena for rotating machinery and combustion and the development of turbulence models. Since 2003, he has been a professor in the Department of Mechanical Engineering at Kumamoto University. He focuses on the production and development of clean energy and renewable energy, thermal fluid flow transport phenomena using nanofluids, advanced cooling device development with the use of nanofluids and development of new clean fuel with the aid of shock-wave.

*Examples of published Guest Editorials are given below for your reference:
https://ietresearch.onlinelibrary.wiley.com/doi/epdf/10.1049/iet-gtd.2020.1493
https://ietresearch.onlinelibrary.wiley.com/doi/epdf/10.1049/iet-pel.2020.0051
https://ietresearch.onlinelibrary.wiley.com/doi/epdf/10.1049/iet-rsn.2020.0089

Contributors

Farha Anjum
Department of ECE
Siddhartha Institute of
　Engineering and Technology
Hyderabad, India

Mohd. Minhajuddin Aquil
School of Engineering
Career Point University
Kota, India

Suresh Ballala
Department of ECE
Sri Indu Institute of Engineering
　and Technology
Hyderabad, India

M. Dharani
Department of ECE
Annamacharya Institute of
　Technology and Sciences
Tirupathi, India

Mir Iqbal Faheem
School of Engineering
Career Point University
Kota, India

Shaik Fairooz
Department of ECE
Malla Reddy Engineering
　College (Autonomous)
Secunderabad, India

Mohammad Illiyas
Department of ECE
Shadan College of Engineering
　and Technology
Hyderabad, India

L. Koteswararao
Department of ECE
KLEF
Hyderabad, India

P. Prasana Murali Krishana
Department of ECE
KITS
Markapur, India

B. Anil Kumar
Department of ECE
GRIET
Hyderabad, India

P. Anil Kumar
CVR College of Engineering
Hyderabad, India

Dayadi Lakshmaiah
Department of ECE
Sri Indu Institute of Engineering
and Technology
Hyderabad, India

I. Satya Narayana
Department of ME
Sri Indu Institute of Engineering
and Technology
Hyderabad, India

S. Pothalaiah
Department of ECE
Vignana Bharathi Institute of
Technology
Hyderabad, India

R. Yadagiri Rao
Department of H&S
Sri Indu Institute of Engineering
and Technology
Hyderabad, India

B. Shilpa
Department of ECE
KLEF
Hyderabad, India

A. Sindhuja
Department of ECE
Sri Indu Institute of Engineering
and Technology
Hyderabad, India

J. B. V. Subrahmnyam
Department of EEE
Dean R&D GIET University
Gunupur, India

T. Venkatakrishnamoorthy
Department of ECE
Sasi Institute of Technology and
Engineering
Tadepalligudem, India

1

A FRAMEWORK FOR IDENTIFICATION OF VEHICULAR TRAFFIC ACCIDENT HOTSPOTS IN COMPLEX NETWORKS

MOHD. MINHAJUDDIN AQUIL AND MIR IQBAL FAHEEM

Career Point University

Contents

DOI: 10.1201/9781003102625-1

1.1 Introduction

Accidents have been a major social problem in developed countries for over 50 years. Since 2001, there has been a growth of 202% of two-wheeler and 286% of four-wheeler vehicles with no road development. Approximately 1.35 million people die each year as a result of road traffic crashes as per a report submitted by the World Health Organization (WHO). The 2030 Schedule for Workable Progress has set an ambitious goal of halving the global number of fatalities and injuries from road traffic crashes by 2020. More than half of all road traffic fatalities are among susceptible road users like pedestrians, cyclists, and riders. It is only in the past decade that developing countries like India have begun to experience a significant increase in the number of road accidents taking place and have found it necessary to institute road safety programs. It is strongly felt that most of the road traffic accidents, being a multi-factor incident, are not only due to a driver's fault on account of the driver's negligence or ignorance of traffic rules and regulations, but also due to many other related parameters such as changes in road geometrics, flow characteristics, road user's behavior, environmental conditions, visibility and absence of traffic guidance, and control and management devices. The Geographical Information System (GIS) emphasizes on providing services on a location scale, and it merely enables the operators to use spatial information and descriptive data to make plans, tables, and diagrams. This system accurately provides search tools; data analysis and results are displayed. The GIS is an organization and decision support system that contains graphic data and site maps that are productive for traffic accident information organization. In the management of road safety, a road traffic accident hotspot is a place where accidents occur frequently. It may have occurred due to various parameters like poor road geometrics, environmental factors, driver's characteristics, and so on. Since few decades, treatment of accident black spot has been the spine of road safety management. In the current scenario, road safety is a major concern. Road safety measures can be adopted by implementing various steps.

1.2 Overview on Vehicular Traffic Accident Hotspot Techniques

To analyze vehicular traffic accident hotspots, various methods have been discussed in this chapter such as Kernel Density Estimation

(KDE), Nearest Neighborhood Hierarchy (NNH), Inverse Distance Weighted (IDW), and Kriging Kim and Levine (1996) described the traffic safety GIS prototype, which performed a spatial analysis of traffic accidents that are developed for Honolulu, Hawaii. Many types of spatial analysis methods based on point, segment, and zone analyses have been developed. Affum and Taylor (1998) introduced a method for traffic management, which is based on a GIS package for studying accident patterns over time. The different methods/techniques are shown in Table 1.1.

1.3 Kernel Density Estimation

KDE is one of the significant spatial analysis tools in the commercially available GIS software package. K divides the entire study area into a pre-determined number of cells. It uses a quadratic kernel function to fit a smoothly elongated surface to each accident location. The surface value reduces from the highest at the incident location point to zero when it reaches a radial distance from the incident location point. The value of the kernel function is assigned to every cell as individual cell values. The resultant density of every cell is computed by adding its i cell values individually. To account for the road accident severity, the weight of each accident is represented as its Identification Number (ID). This facilitates the counting of each accident according to its weight assigned. In case of no injury or severity, according to incident points, the population field is selected as "None". Kernel function can be defined as stated in Eq. (1.1).

$$f(x) = \frac{3}{nh^2\pi} \sum_{i=1}^{n} \left\{ 1 - \frac{1}{h^2} \left[(x - x_i)^2 + (y - y_i)^2 \right] \right\}^2 \quad (1.1)$$

where h is the bandwidth, f is the estimator of the probability density function, π is a constant, x_i and y_i are the deviations of x- and y-coordinates between a point and a known point that is within the bandwidth, and n is the number of known points. The kernel estimator depends upon the choice of bandwidth (h), and hence, suitable bandwidth should be determined according to the purpose of the study. Density values in the raster are predicted values rather than probabilities. It generally gives a smoother surface than the normal estimation method.

Table 1.1 Methods/Techniques for Hotspot Detection

S. NO	TECHNIQUE	FORMULA	PURPOSE
1.	KDE	$f(x) = \dfrac{3}{nb^2\pi} \sum_{i=1}^{n} \left\{ 1 - \dfrac{1}{b^2}\left[(x-x_i)^2 + (y-y_i)^2 \right] \right\}^2$	For smoothing effect within a particular radius and cell size
2.	Point Density	$f(x) = \dfrac{1}{n} \sum_{i=1}^{n} \dfrac{1}{b} w\left(\dfrac{x-x_i}{b} \right)$	Calculates magnitude per unit area using neighborhood operation for a given cell size
3.	Line Density	$f(x) = \dfrac{1}{n} \sum_{i=1}^{n} \dfrac{1}{b} w\left(\dfrac{x-x_i}{b} \right)$	Calculates magnitude per unit area for the radius of the cell size
4.	IDW	$z = \dfrac{\sum_{i=1}^{s} z_i \dfrac{1}{d_i^k}}{\sum_{i=1}^{s} \dfrac{1}{d_i^k}}$	For classifying within the max and min values
5.	Kriging	$f(x) = \sum_{i=1}^{n} \sqrt{i}\,(x^*)\, f\,(x_i)$	For assuming spatial variation of attributes
6.	Spline	$Q(x, y) = \sum A_i d_i^2 \log d_i + a + bx + cy$	For a smoothing effect
7.	Moran's I	$I = \dfrac{\sum_{i=1}^{n} \sum_{i=1}^{n} w_{ij}(x_i - x)(x_j - x)}{s^2 \sum_{i=1}^{n} \sum_{i=1}^{n} w_{ij}}$	For detecting the presence of the clustering of similar values
8.	Getis-Ord GI*	$G(d) = \dfrac{\sum\sum w_{ij}(d)\, x_i x_j}{\sum\sum x_i x_j}\,, \; i \neq j$	For separating the clusters of high and low values

1.4 K-Means Clustering

These hotspots are classified using the clustering process and are organized into classes (or clusters) based on similar attributes. These clusters are then arranged into groups, based on the similarity of the clusters. This hierarchical process allows spatial classification based on the similarity of either characteristic of the accidents within the hotspots or the environmental factors. When determining the database used to build the classification, it is essential to assess the type of data which would be collected and would have the potential of having an impact on accident density.

1.5 Point Density Method

The point density tool estimates the density of point features around each output raster cell. Neighborhood can be shown around each raster cell center, and the number of points that come within it is totaled and divided by the area of the neighborhood. If other than NONE is used in the population field setting, the item's value determines the number of times to count the point. For instance, an item with a value of three would cause the point to be counted as three points. The values can be floating-point or integer. If a unit is selected within the area, then the estimated density for the cell is multiplied by the suitable factor before it is written to the output raster. Although more points will fall inside the broader neighborhood, this number will be divided by a more extensive area when calculating density. It can be evaluated using the equation given below:

$$f(x) = \frac{1}{n} \sum_{i=1}^{n} \frac{1}{h} w\left(\frac{x - x_i}{h}\right) \tag{1.2}$$

The main consequence of a larger radius is that density is calculated considering an additional number of points, which can be farther away from the raster cell. This results in a more generalized output raster as stated in Eq. (1.2), where h is the bandwidth, w is the weight, and n is the number of known points.

1.6 Line Density Method

In this method, the magnitude/unit area from polyline features that lies within a radius around each cell is evaluated. The portion of a line within the neighborhood is considered only when calculating the density. If no lines lie near the neighborhood at a specific cell, that cell is assigned No Data. Larger values of the radius parameter produce a more comprehensive density raster as stated in Eq. (1.3). Lesser values obtained show a raster that has more detail. The output raster values will always be floating points.

$$f(x) = \frac{1}{n} \sum_{i=1}^{n} \frac{1}{h} w\left(\frac{x - x_i}{h}\right) \tag{1.3}$$

where h is the bandwidth, w is the weight, and n is the number of known points.

1.7 Interpolation Method

It is the process in which points with known values are used to estimate values at other unknown points. Interpolation can estimate the accident locations without recorded data by using known accident points at nearby surrounding areas. This type of interpolation is also called a statistical surface. It is the approximate judgment of surface values at points that are unsampled based on the surface values of surrounding points which are known. Interpolation is usually used as a raster operation, but using a TIN (Triangulated Irregular Networks) surface model, it can be used as a vector operation. High cost and limited resources lead to a limited number of selected point locations within that area. In GIS, spatial interpolation of these points can be useful to create a raster surface with estimates made for all raster cells. The output of the interpolation analysis can be used for analyses and modeling of the entire area. Many techniques under the interpolation method can be used such as IDW, Kriging, Spline, and Natural Neighbor.

$$z = \frac{\sum_{i=1}^{s} z_i \frac{1}{d_i^k}}{\sum_{i=1}^{s} \frac{1}{d_i^k}} \tag{1.4}$$

where z_o is the estimated value at point 0, z_i is the value at a known point i, d_i is the distance between point i and point 0, s is the number of known points used in an estimation, and k is the specified power as stated in Eq. (1.4). It is also noticed that this method has some drawbacks like the quality of the interpolation result can decrease if the distribution of sample data points is uneven. Only higher and lower values in that surface can occur at sample data points.

1.8 Kriging Method

The Kriging method involves an interactive investigation of the spatial behavior, unlike interpolation methods that include phenomena characterized by the values, before selecting the best method of estimation for generating the output surface. The distance or direction between points which reflects a spatial correlation was used to describe the difference in the surface. It is a tool that performs mathematical functions to a specified number of points, or all points within a specified radius, to evaluate the output value for each location. This process includes multiple steps, statistical analysis of the data, modeling, creating the surface, and exploring a variance surface. Kriging is most appropriate when you know there is a spatially correlated distance or directional bias in the data that can be simply computed from the observed values, their variance, and the kernel matrix derived from the prior. The formula is given as stated in Eq. (1.5).

$$f(x) = \sum_{i=1}^{n} \sqrt{\lambda_i \left(x^*\right)} f(x_i) \qquad (1.5)$$

In the Kriging method, the weight λ_i depends on a fitted model to the measured points, distance, and spatial relationships among the calculated values around the locations.

1.9 Spline Method

This converts the raster data that allows for local change, gives similar results as an increasing order polynomial transformation, and is also suitable for scanned imagery. It requires only one Google Cloud Platform (GCP). This method is generally preferred over other methods because the interpolation error can be made small even when using lower degree polynomials. This method avoids the problem of Runge's phenomenon. The approximation of thin-plate splines is of the form as stated in Eq. (1.6).

$$Q(x, y) = \sum A_i d_i^2 \log d_i + a + bx + cy \qquad (1.6)$$

where x and y are the x- and y-coordinates of the point to be interpolated, respectively, $d_i^2 = (x - x_i^2)^2 + (y - y_i^2)^2$ and xi and y_i are the x- and y-coordinates of the control point.

1.10 Natural Neighborhood Method

This method was developed by Robin Sibson. This method shows a discrete set of spatial points. Unlike other methods of interpolation, this is a simple method of multivariate interpolation in one or more dimensions. This method selects a value of algorithm of the nearest point and does not consider the values of neighboring points at all. To select color values for a textured surface, an algorithm is used in real-time 3D rendering.

1.11 Mapping Cluster

Mapping cluster also known as spatial autocorrelation is an amount of the degree to which a set of spatial features and data values are associated. It can be clustered, widely scattered, or spaced together. Different methods under mapping cluster or spatial autocorrelation are Anselin Local Moran's I and Getis-Ord GI*.

1.12 Moran's I Method

Moran's I is one of the oldest global spatial autocorrelation indicators which evaluates whether the spatial pattern is clustered, random, or dispersed. It works on both feature locations and feature values simultaneously as stated in Eq. (1.7).

$$I = \frac{\sum_{i=1}^{n} \sum_{i=1}^{n} w_{ij} (x_i - x)(x_j - x)}{s^2 \sum_{i=1}^{n} \sum_{i=1}^{n} w_{ij}} \qquad (1.7)$$

where x_i is the value at point i, x_j is the value at point j, w_{ij} is a coefficient, n is the number of points, and s^2 is the variance of x values with a mean of $\bar{x}$. The coefficient w_{ij} is defined as the inverse of the distance (d) between points i and j or $1/d_{ij}$.

1.13 Getis-Ord GI* Method

This method involves the identification of desired hotspots, collection of events, and mapping of clustering using a spatial database function as stated in Eq. (1.8).

$$G(d) = \frac{\sum\sum w_{ij}(d)x_i x_j}{\sum\sum x_i x_j}, i \neq j \qquad (1.8)$$

where x_i is the value at location i, x_j is the value at location j, and if j is within d of I, $w_{ij}(d)$ is the spatial weight. The weight can be calculated based on some weighted distance. There are distinctions between these methods; a method uses the actual location of an accident (point in GIS) and one that uses the number or density of accidents for a small geographical area such as a grid cell or zone. Getis-Ord GI* is a tool used for performing hotspot analysis. The resultant z-scores and p-values tell you where spatial features lie with either high- or low-value clusters. This application works by looking at each feature within the context of neighboring features. A feature with a high value is exciting but may not be a significant hotspot as shown in Figure 1.1.

A feature will have a high value and is surrounded by other features with high values. The sum of a feature and its neighbors is compared correspondingly to the sum of all features; when the local sum is very different from the expected local sum, and when that difference is too high to be the result of random chance, a statistically significant

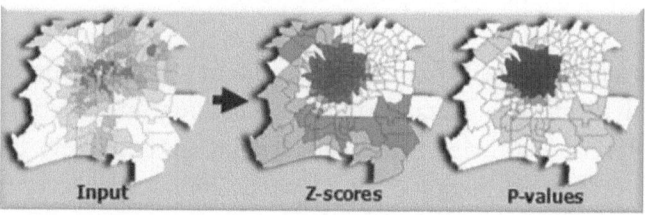

Input   Z-scores   P-values

Figure 1.1   Output of Getis-Ord GI*.

z-score results. FDR correction is applied and eventually statistical significance is adjusted to account for multiple testing and spatial dependency. This method also helps in identifying statistically significant spatial clusters of high values called hotspots and low values called cold spots. For each feature in the input feature class, it creates an output feature class with a z-score, p-value, and confidence level bin field known as a Gi-Bin field. The Gi-Bin field identifies statistically noteworthy hot and cold spots irrespective of whether or not the FDR correction is applied. A higher z-score and a lower p-value for a feature indicate a spatial clustering of high values. A low negative z-score and a small p-value indicate a spatial clustering of lower values. The higher the z-score, the stronger the clustering. A z-score near zero shows no apparent spatial clustering. The z-score is based on the randomization null hypothesis calculation. Getis-Ord GI* is used to add statistical significance to hotspot analysis and for predicting the place where accidents occur.

1.14 Earlier Studies

Anderson (2006) presented a methodology based on KDE using GIS to study the spatial patterns of injury-related road accidents. The study area was London, UK, and the clustering methodology used included environmental data. Results led to a classification of road accident hotspots, environmental data were then added to the hotspot cells, and using K-means clustering, an outcome of similar hotspots was interpreted. The cluster's robustness and potential uses in road safety were assessed and discussed.

Chand et al. (2018) researched accident risk index and accident severity index for different states in India. An index was formed by combining all indices using a set of accident indicators. Values of these two indices have been computed and compared across the states of India.

Kowtanapanich (2007) analyzed accident statistical methods that were used to identify hazardous road locations on the highways in three different countries: Thailand, India, and Turkey. Statistical methods include frequency of accident, accident rate, severity index, quality control, and combined method.

Xie and Yan (2008) used and compared planar and network KDE in terms of pixel size (linear pixel, the same as raster cell, but it is in a network), bandwidth, and density visualization while identifying accident hotspots. However, results show that the KDE network is more appropriate than planar KDE in terms of estimating density estimation for road traffic accidents as planar KDE estimates density beyond accident data context.

Kim et al. (2009) proposed models that can predict road and traffic circumstances, environmental conditions, and driver behaviors during motor vehicle accidents. Accident prediction models attempt to identify potentially dangerous locations of high-frequency motor vehicle accidents on roads. Qualitative measures for road user risk are described as a function of road user characteristics, traffic control devices, and road geometric parameters.

Deepthi and Ganesh (2010) studied the "density" function technique which is available in the Arc GIS software package to identify accident-prone areas in the selected study area. Simple and Kernel densities both were applied in identifying the accident patterns. The road geometric parameters were considered in the accident-prone areas to find out the reasons for the road traffic accident. Suggestions were provided based on the result to reduce road accidents in the future.

Hadayeghi et al. (2010) developed accident models that can illustrate the relationship between accident frequency and a variety of additional variables such as traffic volume, roadway characteristics, or socioeconomic and demographic features.

Wang C (2010) attempted to establish the relationship between traffic congestion and road accidents by using an econometric and GIS approach. The data used were between 2003 and 2007 data from the M25 motorway and its adjacent roads. The UK adopts a speed limit in the roadway for reducing traffic congestion and accidents; this also prospers a new map-matching technique for road segments and frequent accidents and then identifies the hazardous accident hotspots for road safety and control measures.

Dai et al. (2010) researched on GIS, which is one of the tools used to analyze crashes for which crash information has to be converted meaningfully for spatial mapping. Information regarding the number of crashes, time of crashes, and locations was taken. By using this information, crash data were analyzed spatially using GIS, and the high crash risk areas (hotspots) and the times at which most crashes occurred were found by this method.

Steen Berghen et al. (2010) researched on GIS, which is one of the tools used to analyze crashes for which crash information has to be converted meaningfully for spatial mapping. For such a conversion, information regarding the number of crashes, time of crashes, and locations is needed. By using this information, crash data can be analyzed spatially using GIS.

Prasanna Kumar et al. (2011) researched on road accident hotspots in a South Indian city. Road traffic accident and hotspot spatial densities were determined using the Moran's I method of spatial autocorrelation, Getis-Ord GI statistics, and point Kernel density functions. The traffic accidents show a clustered nature while comparing with spatio-temporal breakups and show random distribution in certain modules. The results of hotspot analysis can be estimated using Kernel density, which delineates the road stretches as well as isolated zones where the hotspots are concentrated.

Plug et al. (2011) researched on GIS. It can provide us information on the high crash risk areas (hotspots) and the times that most crashes occurred. To identify clusters of various road traffic incidents, a number of studies successfully implemented GIS techniques.

Rodzi et al. (2015) researched on how to reduce the number of road accidents, especially in Malaysia since it could be a big threat to this country. Due to the lack of a complete road traffic accident recording and analysis system, it can be effective. By using the : Infrared Astronomy Satellite (IRAS) system, the traffic police would be able to control and manage whole traffic accidents in a real-time monitoring system. For road traffic analysis and management, a GIS-based solution can be used for comprehensive intelligence.

Chang et al. (2011) proposed a fuzzy K-means clustering algorithm by extending the K-means clustering algorithm with fuzzy rules. In their model, the authors used center displacement for making cluster decisions. However, the distance-based clustering methods perform better than clustering using center displacement. Therefore, it is necessary to propose a fuzzy clustering algorithm based on distance.

Parasanna Kumar et al. (2011) researched on clustering analysis with respect to accident type and occurrence time, using Moran's I index to perform spatial autocorrelation. Clustering analysis was performed using KDE and Getis-Ord GI* statistics. Road traffic accident hot and cold spots were determined by using the clustering analysis method. From this study, high speed was found to be an indicator factor of single-vehicle crashes (run-off crashes).

Eckley and Curtin (2013) researched on clustering of traffic accidents on a road network using Sanet software and studied road traffic accident clusters within the boundary of the selected study area. Traffic accidents were determined for different seasons, days, and time periods. KDE, K-means clustering, and NNH clustering are used to show road traffic accident hotspots. Zones with hotspots were identified according to seasons, day, and time periods in terms of incidence. Due to the lack of data, only South Anatolian Motorway was analyzed in detail numerically. In the study area, most accident types identified were rear-ended, colliding with stationary objects, and run-off.

Nagarajan et al. (2012) researched on the use of remote sensing (RS) and GIS for the identification of black spots and accident analysis for a particular stretch of NH-45 starting from Tambaram to Chengalpet. There are 11 traffic accident locations identified in this study using Arc GIS software and a high-resolution satellite map was developed based on the data collected from the traffic police department, field survey conducted for traffic volume, vehicle spot speed, and plotting of the study area.

Singh et al. (2012) developed an accident prediction model based on Annual Average Daily Traffic (AADT) and road conditions and studied the effect of traffic volume on road accident rates. NH-77 was selected as the study area that passes through the cities of Hajipur and Muzaffarpur and data were collected using statistical methods for prediction of road traffic accidents. The estimated values from the accident prediction model were tested by a chi-squared test.

Apparao et al. (2013) researched on accident analysis statistical methods (accident frequency, accident rate, severity index, quality control, and combined method), which were used to identify hazardous road locations on the highways in Thailand.

Liyamol et al. (2013) researched on the use of GIS for identification of accident black spots. Accidents can occur at any instance of time with multi-factors depending on the situation including parameters like driver characteristics, road user characteristics, and environmental characteristics. The study carried out by the Kerala Road Safety Authority found that the maximum numbers of accident-prone stretches or the black spots are in the districts of Alappuzha and Ernakulam.

Blazquez CA et al. (2013) researched on Moran's I index test, which is used to identify a positive spatial autocorrelation on accident contributing factors, times of day, straight road sections, and intersections.

Parek et al. (2015) conducted accident analysis and identification of black spots. The objective was to analyze road traffic accident data and identify black spots. Finally, they concluded the following estimations from accident analysis: a maximum number of accidents occur due to head collision when there was no facility median on the center of the lane. In this study, they concluded that the maximum number of road accidents occur in March and April. The majority of accidents occur in the summer season.

Patel Savan Kumar et al. (2014) carried out an analysis of road accidents, which aimed to analyze the traffic accidents occurring in a selected stretch because of the vast number of road user incidences of clients; in particular, four-wheelers were a concern. Road traffic accident data were collected from the department of police of the study area. The collected data were analyzed according to parameters such as yearly variation of the accident, classified according to the month, according to the day, according to the collision type, according to the accident spot, and so on, and the authors concluded their study by proposing safety measures.

Gupta et al. (2014) attempted to study accident black spot validation using GIS in Chandigarh. The promising idea of this research is the utilization of GIS tools for the study. The use of ArcGIS tools for accident mapping facilitated the easy recognition of accident spots. The spatial autocorrelation tools aided to derive more accurate results in identifying the significant hotspots termed as black spots, which are the major risk zones that seek immediate attention on mitigation measures to prevent frequent accidents.

Saleh et al. (2010) carried out a GIS-based research to identify road traffic accident black spots in the Federal Capital Territory of Nigeria. A complete database of road traffic and accident data and other relevant information was developed, which will be needful and useful for Nigeria Police, Federal Road Safety Commission, and Road Traffic Services, and other stakeholders.

Vyas, P.R. et al. (2012) made an effort on State Highway, Karnataka to identify the significant hotspots, i.e., where frequent accidents occur, which are also termed as black spots.

Kumar et al. (2014) researched on a case study on Vishakhapatnam city; their work was intended to identify accident spots in the study area by using collected police record data and the optimum route was identified between accident spots and hospitals using the Arc GIS-10.2 analysis tool.

Yakar (2015) carried out research on road traffic accidents, studied about the parameters related to the road and environment, and demonstrated that future traffic accidents will occur under the same conditions as past traffic accidents. This study investigates the use of the frequency ratio method in the determination of accident-prone road stretch. This research aimed to develop a relationship between accident number and environmental properties of road sections and also identified accident-prone road sections.

Sorate et al. (2015) studied National Highway-4 (from Katraj Tunnel to Chandni Chowk) for the identification of black spots on the selected stretch as a potential black spot. This study included several analytical methods to identify the black spots such as ranking method and severity index, density method, and weighted severity index to process the primary and secondary data. A GIS-based weighted severity index method was used in this study to identify such black spots.

Subba Reddy et al. (2019) studied accident black spots using mixed traffic sheets in developed cities. The authors also suggested suitable remedial measures to develop an accident prediction model, which can reduce the life losses and injuries in the study area.

Bobade et al. (2015) attempted a study on black spots identification on national highways and expressways and stated that around 13 people die per day due to road traffic accidents. In this study, an experimental investigation was undertaken on the past accident data and they compared the data with the data recorded in police records. Finally, they derived ten parameters and used a ranking method to identify the accident black spots.

Thakali et al. (2015) researched on the identification of crash spots on a road network. This study aimed at a comparison of two geostatistical-based methods named KDE and Kriging. It was found that based on the Personnel Asset Inventory (PAI) measure, the Kriging method outperformed as compared to the KDE method in the detection of accident hotspots, with respect to the groups of crash data at different times of the day. It was found that the results were completely different in both methods, hence signifying the importance of the selection of the right geostatistical method for identification of road accident hotspots.

Anitha and Arulraj (2016) researched on the present state of traffic accident information on NH-47 from Gandhipuram to Avinashi, Coimbatore District and also discussed the identification of high-risk accident locations by using the GIS tool and the lack of safety on selected study areas on the highway. Remedial measures and provisions for traffic safety were suggested for reducing the risk of accidents in black spots.

Romano et al. (2017) studied numerous techniques to identify spatial accident hotspots on the road network using KDE and significant linear route detection, but the time dimension and temporal dynamics of hotspots are not considered. Limitations were demonstrated using existing methods and a new method called spatial-temporal network KDE that involves these features.

Chandrasekar et al. (2017) researched on the identification of black spots (high-risk accident locations, accident-prone locations) in the Puttur to Ramagiri stretch. Accident data consisted of time of incidence, kind of collision, type of vehicles, victims' details, and persons injured. Using the data, ten black spots were identified and engineering surveys were conducted to know the road characteristics. Parameters such as the width of the road, alignment of the road, number of aspect roads, and traffic volume were considered.

Sarifah et al. (2018) researched on suitable techniques for the determination of hotspots on Malaysian highways. This study used spatial analysis techniques, namely, nearest neighborhood hierarchical clustering and spatial-temporal clustering, using CrimeStat and visualizing in the ArcGIS™ tool to estimate the concentration of the traffic incidents. Results were compared based on their accuracies, identified numerous hotspots, and showed that they varied in number with places, depending on the parameters. Further analysis showed that the spatial-temporal clustering has a higher accuracy index compared to the nearest neighbor hierarchical (NNH) clustering on the selected stretch.

Colak et al. (2018) identified the intensity of traffic accidents (hotspot regions) using spatial statistics techniques. The GIS tool was used to examine the hotspots for traffic safety. This study used analysis based on network spatial weights to identify black spots. Hotspot analysis was carried out using Getis-Ord GI* and road accident data were used to generate network spatial weights. Then, the Kernel density method was used to define traffic accident black spots.

Faheem et al. (2019) identified the accident black spots in Hyderabad city and developed an accident prediction model using the multiple linear regression method. They developed the prediction models by considering the various accident-causing parameters. QGIS software has been used for the study, which included a collection of spatial data and non-spatial data for the identification of traffic accident black spots. The required data were then collected (zone-wise), which is comprised of accidents caused due to various factors. Different equations were developed for different zones, which help in the prediction of accidents. They suggested necessary improvements to be made at the black spot locations to reduce traffic accidents.

1.15 Research Directions

The majority of the existing works carried on the identification of vehicular traffic accident hotspot methods considered the road traffic parameters individually or in combination. This chapter provides a framework of different methods for the identification of vehicular traffic accident hotspots. Many authors developed different models out of which maximum work has been carried out on KDE, Kriging, and NNH methods, whereas minimum work has been carried out on spline, point density, line density, IDW, and Moran's I methods. After studying the different methods stated above, it is found that least work has been carried out on the Getis-Ord GI* method.

1.16 Research Framework

A research framework for the identification of vehicular traffic accident hotspots in complex networks is shown in Figure 1.2.

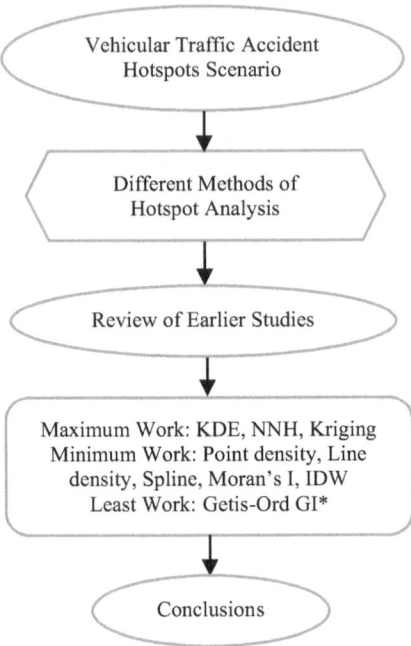

Figure 1.2 Research framework for the identification of vehicular road traffic accident hotspots.

This chapter attempts to explore the future research trend with respect to the literature survey done so far, which is characterized as the aim and objective of the study, to separate the clusters of high and low values, to apply for point and polygon features using Getis-Ord GI*, to limit the traffic congestion and ease the number of accidents by investigating into the chronology of the accidents using GIS methods, and to provide recommendation for ensuring the road user's safety.

1.17 Conclusions

This chapter attempts to study the united views of different researchers on the identification of vehicular road accident hotspots. This research provides a lot in insight about the conditions of road accidents. This research establishes to be precious in the identification of a large quantity of vehicular road accident hotspots. The GIS method shall be employed to evaluate vehicular road accidents in complex networks.

References

Anderson T, (2006), "Comparison of spatial methods for measuring road accident hotspots: A case study of London", *Journal of Maps*, ISSN 1744-5647, Vol. 3, No. 1, January 2006, pp 55–63.

Anderson TK, (2009), "Kernel density estimation and K-means clustering to profile road accident hotspots", *Accident Analysis and Prevention*, ISSN 0001-4575/$, Vol. 41, No. 3, May 2009, pp 359–364.

Anitha S & P Arulraj, (2016), "Identification of hotspots of traffic accidents using GIS", *International Journal of Advances in Engineering and Technology*, E-ISSN 0976-3945, Vol. 7, No. 3, September 2016, pp 13–16.

Apparo, G., Millikarjunareddy, P., & Raju, S.G. (2013). "Identification of Accident Black Spots for National Highway Using GIS." *International Journal of Scientific and Technology Research*, Vol. 2, No. 2, pp. 154–157.

Azad A, (2017), "Road crash prediction models: Different statistical modelling approaches", *Journal of transportation technologies*, ISSN 2160-0481, Vol. 7, April 2017, pp 190–205.

Blazquez, C. A., & Celis, M. S. (2013). "A spatial and temporal analysis of child pedestrian crashes in Santiago, Chile." *Accident Analysis & Prevention*, Vol. 50, pp. 304–311.

Bobade, S. U., Sorate, R. R., Kulkarni, R. P., Patil, M. S., Talathi, A. M., Sayyad, I. Y., & Apte, S. V. (2015). "Identification of accident black spots on national highway 4 (New Katraj tunnel to Chandani chowk)." *IOSR Journal of Mechanical and Civil Engineering (IOSR-JMCE)*, Vol. 12, No. 3, pp. 61–67.

Chand, S., & Dixit, V. V. (2018). "Application of Fractal theory for crash rate prediction: Insights from random parameters and latent class tobit models." *Accident Analysis & Prevention*, Vol. 112, pp. 30–38.

Chandrasekhar R et al., (2017), "Study of accidents on NH-140 and its preventive measures", *International Journal of Advance Engineering and Research Development*, E-ISSN: 2348-4470, Vol. 4, No. 8, August 2017, pp 133–137.

Chang, C. T., Lai, J. Z., & Jeng, M. D. (2011). "A fuzzy k-means clustering algorithm using cluster center displacement." *J. Inf. Sci. Eng.*, Vol. 27, No. 3, pp. 995–1009.

Colak, H. E., Memisoglu, T., Erbas, Y. S., & Bediroglu, S. (2018). "Hot spot analysis based on network spatial weights to determine spatial statistics of traffic accidents in Rize, Turkey. Arabian Journal of Geosciences," Vol. 11, No. 7, pp. 1–11.

Dai, C. Q., Zhu, S. Q., Wang, L. L., & Zhang, J. F., (2010), "Exact spatial similaritons for the generalized (2+ 1)-dimensional nonlinear Schrödinger equation with distributed coefficients", *EPL (Europhysics Letters)*, Vol. 92, No. 2, p 24005.

Deepthi JK & B Ganesh, (2010), "Identification of accident hot spots: A GIS based Implementation on Kunnur District, Kerala", *International Journal of Genetics and Genomics*, ISSN 0976-4380, Vol. 1, No. 1, January 2010, pp 51–59.

Eckley, D. C., & Curtin, K. M., (2013), "Evaluating the spatiotemporal clustering of traffic incidents", *Computers, Environment and Urban Systems*, Vol. 37, pp 70–81.

Faheem MI et al., (2018), "A GIS approach road infrastructure safety management using GIS", *International Journal of Technical Innovation in Modern Engineering and Science*, E-ISSN 2455-2585, Vol. 4, No. 8, August 2018, 250–261.

Fattibene, P., Wieser, A., Adolfsson, E., Benevides, L. A., Brai, M., Callens, F., ... Zhumadilov, K., (2011), "The 4th international comparison on EPR dosimetry with tooth enamel: Part 1: Report on the results", *Radiation Measurements*, Vol. 46, No. 9, pp 765–771.

Faheem MI et al., (2019), "Identification of blackspots and development of prediction model using multiple linear regression model", *International Journal of Research and Analytical Reviews*, E-ISSN 2348-1269, Vol. 6, No. 2, June 2019, pp 680–687.

Gebretensay FB & J Juremalani et al., (2018), "Road traffic accident analysis and prediction model: a case study of Vadodara city", *International Research Journal of Engineering and Technology*, ISSN 2395-0056, Vol. 5, No. 1, January 2018, pp 191–196.

Gupta, A. K., Sharma, S., & Redhu, P. (2014). "Analyses of lattice traffic flow model on a gradient highway." *Communications in Theoretical Physics*, Vol. 62, No. 3, p. 393.

Hadayeghi, A., Shalaby, A., & Persaud, B., (2010), "Development of planning-level transportation safety models using full Bayesian semiparametric additive techniques", *Journal of Transportation Safety & Security*, Vol. 2, No. 1, pp 45–68.

Harinath, Y., Reddy, D. H. K., Kumar, B. N., Apparao, C., & Seshaiah, K., (2013), "Synthesis, spectral characterization and antioxidant activity studies of a bidentate Schiff base, 5-methyl thiophene-2-carboxaldehyde-carbohydrazone and its Cd (II), Cu (II), Ni (II) and Zn (II) complexes", *Spectrochimica Acta Part A: Molecular and Biomolecular Spectroscopy*, Vol. 101, pp 264–272.

Holmlund, H., Lindahl, M., & Plug, E., (2011), "The causal effect of parents' schooling on children's schooling: A comparison of estimation methods", *Journal of Economic Literature*, Vol. 49, No. 3, pp 615–651.

Kim, W., Mohrig, D., Twilley, R., Paola, C., & Parker, G., (2009), "Is it feasible to build new land in the Mississippi river delta?", *Eos, Transactions American Geophysical Union*, Vol. 90, No. 42, pp 373–374.

Kiran N et al., (2017), "Prediction of road accident modelling for Indian national highways", *International Journal of Civil Engineering and Technology*, E-ISSN 0976–6316, Vol. 8, No. 1, January 2017, pp 789–802.

Kowtanapanich, W., Tanaboriboon, Y., & Charnkol, T. (2007). "A Prototype Of The GIS-Based Traffic Accident Database System: Thailand Case Study." *Journal of the Eastern Asia Society for Transportation Studies*, Vol. 7, pp. 2757–2769.

Kumar, P., McElhinney, C. P., Lewis, P., & McCarthy, T. (2014). "Automated road markings extraction from mobile laser scanning data." *International Journal of Applied Earth Observation and Geoinformation*, Vol. 32, pp. 125–137.

Kumar M et al., (2017), "Identification of accident black spot locations using GIS & GPS technology: A case study of Hyderabad", *International Journal of Innovative Research in Science, Engineering and Technology*, ISSN 2319-8753, Vol. 6, No. 1, January 2017, pp 809–817.

Kumar, D., Pal, A., Kumar, A., & Ahirwar, V., (2014), "Thermally stimulated discharge current study of PMMA: PVP blends", *International Journal of Pharmaceutical Science Invention*, Vol. 3, pp 44–50.

Lewis, N. E., Nagarajan, H., & Palsson, B. O., (2012), "Constraining the metabolic genotype–phenotype relationship using a phylogeny of in silico methods", *Nature Reviews Microbiology*, Vol. 10, No. 4, pp 291–305.

Liyamol, Isen, Shibu, A., & Saran, M.S., (2013), "Evaluation and treatment of accident black spots using Geographic Information System", *International Journal of Innovative Research in Science, Engineering and Technology*, Vol. 2, No. 8, pp 3865–3873.

Martin J, (2002), "Relationship between crash rate and hourly traffic flow on interurban motorways" *Accident Analysis and Prevention*, ISSN 0001-4575, Vol. 34, No. 5, September 2002, pp 619–629.

Meshram K et al., (2013), "Accident analysis on national highway-3 between Indore to Dhamnod", *International Journal of Application or Innovation in Engineering & Management*, ISSN 2319-4847, Vol. 2, No. 7, July 2013, pp 57–59.

Mohan A & VS Landge, (2017), "Identification of accident black spots on national highway" *International Journal of Civil Engineering and Technology*, ISSN 0976-6316, Vol. 8, No. 4, April 2017, pp 588–596.

Nagarajan, R., Thirumalaisamy, S., & Lakshumanan, E. (2012). "Impact of leachate on groundwater pollution due to non-engineered municipal solid waste landfill sites of erode city, Tamil Nadu, India." *Iranian journal of environmental health science & engineering*, Vol. 9, No. 1, pp. 1–12.

Pandey RS & S Dhobale, (2017), "Identification and improvement of accident black spots on N.H.86 district Sagar-Madhya Pradesh", *International Research Journal of Engineering and Technology*, E-ISSN 2395-0056, Vol. 4 No. 9, September 2017, pp 740–742.

Parekh, V. P., Labana, A. B., & Parikh, V. A. (2015). "Literature review on road accident analysis a case study on Dahod to Jhalod section of NH 113."

Patel, A. B., Bhatt, N. K., Thakore, B. Y., Vyas, P. R., & Jani, A. R., (2014), "The temperature-dependent electrical transport properties of liquid Sn using pseudopotential theory", *Molecular Physics*, Vol. 112, No. 15, pp 2000–2004.

Patel, S. H., Baj, Y. M., & Patel, N. K., (2018), "Synthesis, characterization and optimization of reaction parameters for sodium salt of partially carboxymethylated okra gum", *Journal of Applied Polymer Science*, Vol. 7, No. 1, pp 55–62.

Patel, S. K., & Kumar, A. (2014). "EAANS: Energy aware adjacent node selection based optimal shortest path detection in wireless sensor network." *International Journal of Emerging Technology and Advanced Engineering.*

Plug, C., Xia, J. C., & Caulfield, C. (2011). "Spatial and temporal visualisation techniques for crash analysis. Accident Analysis & Prevention," Vol. 43, No. 6, pp. 1937–1946.

Prasannakumar, V., Vijith, H., Charutha, R., & Geetha, N. (2011). "Spatio-temporal clustering of road accidents: GIS based analysis and assessment." *Procedia-social and behavioral sciences*, Vol. 21, pp. 317–325.

Reddy, G. R. S. (2019). "A taxonomy of issues, challenges and applications in internet of multimedia things (IoMMT)." *i-manager's Journal on Cloud Computing*, Vol. 6, No. 1, p. 1.

Reshma EK & SU Sharif, (2012), "Prioritization of accident black spots using GIS", *International Journal of Emerging Technology and Advanced Engineering*, ISSN 2250-2459, Vol. 2, No. 9, September 2012, pp 117–122.

Rodzi, N. A. H. M., Othman, M. S., & Yusuf, L. M. (2015, October). "Significance of data integration and ETL in business intelligence framework for higher education." In *2015 International Conference on Science in Information Technology (ICSITech), IEEE*, pp. 181–186.

Romano, B., & Jiang, Z., (2017, November), "Visualizing traffic accident hotspots based on spatial-temporal network kernel density estimation", In *Proceedings of the 25th ACM SIGSPATIAL International Conference on Advances in Geographic Information Systems* (pp. 1–4).

Romano, B. (2017, March). "Managing the internet of things." In *Proceedings of the 2017 ACM SIGCSE Technical Symposium on Computer Science Education.* pp. 777–778.

Saleh, R., Marks, M., Heo, J., Adams, P. J., Donahue, N. M., & Robinson, A. L., (2015), "Contribution of brown carbon and lensing to the direct radiative effect of carbonaceous aerosols from biomass and biofuel burning emissions", *Journal of Geophysical Research: Atmospheres*, Vol. 120, No. 19, pp 10–285.

Saleh, W., Kumar, R., & Sharma, A. (2010). "Driving cycle for motorcycles in modern cities: case studies of Edinburgh and Delhi." *World Journal of Science, Technology and Sustainable Development*, Vol. 7, No. 3, pp. 263–274. https://doi.org/10.1108/20425945201000017.

Sarifah SSR et al., (2018), "Determining hotspots of road accidents using spatial analysis" *International Journal of Research in Advent Technology*, ISSN 2502-4752, Vol. 9, No. 1, January 2018, pp 146–151.

Singh, R. K., & Suman, S. K. (2012). "Accident analysis and prediction of model on national highways. *International journal of Advanced technology in civil Engineering*," Vol. 1, No. 2, pp. 25–30.

Sorate, R. R., Kulkarni, R. P., Bobade, S. U., Patil, M. S., Talathi, A. M., Sayyad, I. Y., & Apte, S. V. (2015). "Identification of accident black spots on national highway 4 (New Katraj tunnel to Chandani chowk)." *IOSR Journal of Mechanical and Civil Engineering (IOSR-JMCE)*, Vol. 12, No. 3, pp. 61–67.

Steenberghen, T., Aerts, K., & Thomas, I., (2010), "Spatial clustering of events on a network", *Journal of Transport Geography*, Vol. 18, No. 3, pp 411–418.

Thakali, L., Kwon, T. J., & Fu, L., (2015), "Identification of crash hotspots using kernel density estimation and kriging methods: A comparison", *Journal of Modern Transportation*, Vol. 23, No. 2, pp 93–106.

Tong, K. M., Chen, C. P., Huang, K. C., Shieh, D. C., Cheng, H. C., Tzeng, C. Y., ... Tang, C. H., (2011), "Adiponectin increases MMP-3 expression in human chondrocytes through AdipoR1 signaling pathway", *Journal of cellular biochemistry*, Vol. 112, No. 5, pp 1431–1440.

Uzun, Y., Kaya, A., Karacan, İ. H., Kaya, Ö. F., & Yakar, S., (2015), "Macromycetes determined in Islahiye (Gaziantep/Turkey) district", *Biological Diversity and Conservation*, Vol. 8, pp 209–217.

Vyas, P. A. V. A. N., Honnappanavar, M. L., & Balakrishna, H. B. (2012). "Identification of black spots for safe commuting using weighted severity index and GIS. people," Vol. 31, No. 2.

Wang, C. (2010). *The relationship between traffic congestion and road accidents: an econometric approach using GIS* (Doctoral dissertation, © Chao Wang).

Xie Z & J Yan et al., (2008), "Kernel density estimation of traffic accidents in a network space" *Computers, Environment and Urban Systems*, ISSN 0198-9715/$, Vol. 32, No. 5, September 2008, pp 396–406.

Yahya, M., Whang, S., Gupta, R., & Halevy, A., (2014, October), "Renoun: Fact extraction for nominal attributes", In *Proceedings of the 2014 Conference on Empirical Methods in Natural Language Processing (EMNLP)* (pp. 325–335).

Yakar, F. (2015). "Identification of accident-prone road sections by using relative frequency method." *Promet-Traffic&Transportation*, Vol. 27, No. 6, pp. 539–547.

Yildirim, V., Memisoglu, T., Bediroglu, S., & Colak, H. E., (2018), "Municipal solid waste landfill site selection using multi-criteria decision making and GIS: Case study of Bursa province", *Journal of Environmental Engineering and Landscape Management*, Vol. 26, No. 2, pp 107–119.

2

TRAFFIC SAFETY MANAGEMENT USING SPATIAL ANALYSIS AND CLUSTERING METHODS

MOHD. MINHAJUDDIN AQUIL AND MIR IQBAL FAHEEM

Career Point University

Contents

2.1 Introduction

India has the second-largest road network in the world with 3,300,000 km of connectivity. In India, there has been a growth of vehicles at an average pace of 10.16% per annum over the last five years (NHAI). Road accident statistics for the last ten years shows a steep rise in both the number of accidents and the number of fatalities. Experience, as well as literature on crash identification, indicates

DOI: 10.1201/9781003102625-2

that not only the frequencies of accidents but also the exposure is important in quantifying the risk of a location. The road geometry parameters can also be considered in accident-prone areas to find out the reasons for the accidents in the geographical space for identifying the high density of road traffic accidents (Faheem et al., 2017). Detection of accident hotspots is a process of identification of road segments that have a high risk of road traffic accidents (Thakali et al., 2014). For this purpose, road traffic accident data, road traffic volume, and other related information are required. Geographical Information Systems (GIS)-aided spatial analysis enables traffic safety organizations to decrease the traffic accidents at high-risk sites, subsequently minimizing road traffic accidents yielding to the degree of physical injuries. But the success of these analysis methods depends widely on the road traffic accident data like precision, reliability, and comprehensiveness, which are very important for inputting data into GIS and performing spatial analysis for road traffic safety enhancement (Erdogan et al., 2008). By identifying road traffic accident hotspots using GIS along with value-added data, a more vigorous understanding can be achieved for road traffic safety improvements (Anderson, 2009). Additionally, GIS combined with complex statistical analysis and intelligence tools helps us to understand the causes of accidents at those locations and find different ways of reducing them. This research paper is arranged as follows: firstly, a review of the existing literature is presented followed by a study area description. It is then followed by research methodology with analysis procedure and then finally, results are concluded.

2.2 Earlier Studies

In this section, the previous work done on the identification of accident hotspots using different methods and models with the overview of the relevant research works is discussed. Faheem et al. (2019) identified critical spatial and non-spatial data for the identification of road traffic accident black spots. Zone-wise data were collected which is comprised of accidents caused due to several parameters followed by the development of an accident prediction model. Also, different equations were developed for different zones, which helped in the prediction of accidents. Finally, the study suggested making necessary improvements to be made at the incident locations to reduce road traffic accidents. Gebretensay et al. (2018) developed road traffic accident prediction models based on different road parameters. Pandey et al. (2017) investigated road accidents on the National Highway-86, Sagar to Shahgarh in district Sagar (M.P.), India. Weighted severity index (WSI) was used to identify the road traffic accident black spots. Furthermore, severity scores were assigned based on the number and degree of accident severity in that particular location in the last few years using the WSI method.

The advantage of using GIS that allows analysis which is not possible with the other database management systems in identifying road traffic accident hotspots is its ability to access and analyze spatial data with respect to its actual spatial location of the area. Several hotspot analysis methods are used for detecting traffic accident hotspots in GIS environments such as spatial analysis methods, interpolation methods, and mapping cluster. Thakali et al., (2015) compared two geo-statistical-based methods, namely Kernel Density Estimation (KDE) and Kriging for identifying crash hotspots on a selected road network. The study indicated that the Kriging method outperformed the KDE method in its ability to delineate traffic road accident hotspots. Additionally, Erdogan et al., (2008) also carried out accident analysis using the KDE method and repeatability analysis. The study aimed at identifying high rate accident locations and safety deficient areas on the highways of the city of Afyonkarahisar. Faheem et al. (2018) investigated the development of a processing tool to create black spot analysis by the KDE method using GIS to reduce traffic accidents. The study suggested improvements that can be implemented in order to have a more user-friendly and automated system and to make data accessible for all the road users and to implement various measures to provide more safety on roads.

Anderson (2009) identified road accident hotspots in London, UK. KDE and K-means clustering were performed to locate similar hotspots. Five groups and 15 clusters were created based on collision and results were discussed on the same. Selvasofia (2016) discussed the scenario of a road traffic accident on NH-47 Gandhipuram to Avinashi and NH-209 from Gandhipuram to Annur, Coimbatore district. The GIS tool was used for the analysis and remedial measures were discussed and suggestions were made for reducing the intensity of accidents black spots. Blazquez et al. (2013) presented a spatial and temporal analysis of child pedestrian crash data in Santiago, Chile. KDE was applied and clusters were determined using the GIS tool. Moran's I index test was also performed, which identified positive spatial autocorrelation on crash contributing factors, time of day, and so on; within the critical areas obtained, there was no statistical significance with respect to gender, weekday, and month of the year. Bhagyaiah et al. (2014) discussed the road accident situations at the selected stretches of Hyderabad city and discussed the state of crashes yielding to the injuries. Sarifah et al. (2018) analyzed two spatial techniques, namely, nearest neighborhood hierarchy (NNH) clustering and spatial-temporal clustering, using the CrimeStat software. The Arc GIS™ tool was used to visualize the incidents and the results were compared based on their accuracies. The results identified several hotspots and showed that they varied in number and locations, depending on different road parameters. Prasannakumar et al., 2011 assessed spatial clustering of accident and hotspot spatial densities using Moran's I method, Getis-Ord Gi* statistics, and point Kernel density functions. The Kernel density surface, projected from the results of hotspot analysis, describes the road stretches as well as isolated zones where the hotspots are concentrated. The literature review focuses on many different methodologies and criteria that have been developed for improving the accuracy of the hotspot identification process.

It can be observed from the literature that among several methods of hotspot analysis, the KDE method is the most widely used in the GIS environment as this method is simple and easy to implement. Also, least work has been carried out using Getis-Ord Gi* and NNH.

2.3 Accident Hotspot Methods—An Overview

No previous research has analyzed road traffic accident hotspots using these methods in the GIS environment in the selected case study area. An outline of the research methodology for the intended study is presented in Figure 2.4. As mentioned previously, the two methods which are employed in this research are Getis-Ord Gi* and NNH, which are discussed below.

2.4 Getis-Ord Gi* Method

The two key processes involved in the identification of desired hotspots are a collection of events and mapping of clustering using the Getis-Ord Gi* function as stated in Eq. (2.1).

$$G(d) = \frac{\sum\sum w_{ij}(d)x_i x_j}{\sum\sum x_i x_j} \; i \neq j \; G(d) = \frac{\sum\sum w_{ij}(d)x_i x_j}{\sum\sum x_i x_j}, i \neq j$$

(2.1)

where x_i is the value at location i, x_j is the value at location j, and if j is within $d(i)$, and $w_{ij}(d)$ is the spatial weight. The weights are often supported by some weighted distance. There are distinctions between these methods; a method uses the actual location of an accident (point in GIS) and one that uses the number or density of accidents for a small geographical area like a zone or a grid cell. The hotspot analysis is a method that computes the Getis-Ord Gi* statistics in a dataset for each feature. The resultant z-scores and p-values tell you where spatial features lie with either high or low value clusters. This application works by looking at each feature within the context of neighboring features. A feature with a high value is exciting but may not be a significant hotspot score and value as shown in Figure 2.1. This method helps in identifying strategically significant spatial clusters of high values as hotspots and of low values as cold spots.

This also creates an output feature class with a z-score and p-value for every feature within the input feature class. The z-score and p-value fields do not reflect any sort of FDR correction. The Gi-Bin field identifies statistically significant hot and cold spots, no matter whether or not the FDR correction is applied. A high z-score and a less p-value specify a spatial clustering of low values. The higher

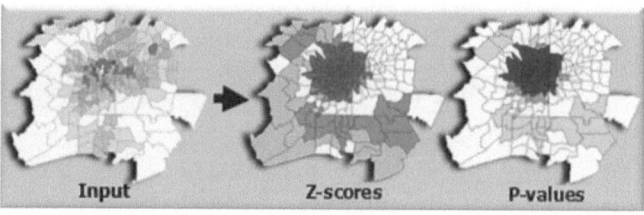

Figure 2.1 Hotspot scores and values.

(or lower) the z-score, the stronger the clustering. The z-score is predicted on the randomization null hypothesis calculation. Getis–Ord Gi^* is employed to feature the statistical significance of hotspot analysis and for predicting where accidents occur.

2.5 Nearest Neighborhood Hierarchical Method

In this study, NNH clustering is obtained using the CrimeStat software that outputs clusters in two formats, i.e., convex hulls and ellipses. The user can visualize the clusters in these two different formats. There are both advantages and disadvantages for each format. Convex hulls are polygons that cover exactly all the clustered points. For a detailed analysis, convex hulls are preferred to the ellipses since the convex hull is more precise and defines the actual area where the hotspot occurs. Ellipse is more like a symbolic representation of the cluster, which generally looks better on a map and is easily understood by users. As can be seen in Figure 2.2, the ellipses can stick out beyond the actual locations but also can cut off part of the hotspot area where they are a mathematical abstraction (Levine, 2010).

2.6 CrimeStat Software

CrimeStat is a spatial statistical program that interfaces with desktop GIS packages for visualization. It is a stand-alone Windows program for the analysis of crime incident locations that can also be used for locating traffic accident hotspots and can interface with desktop GIS programs as shown in Figure 2.3. The program writes calculated

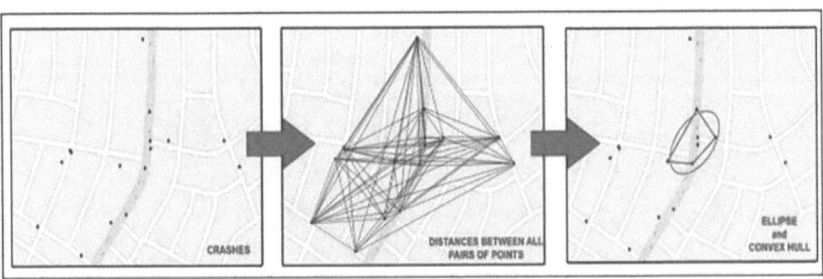

Figure 2.2   Clustering.

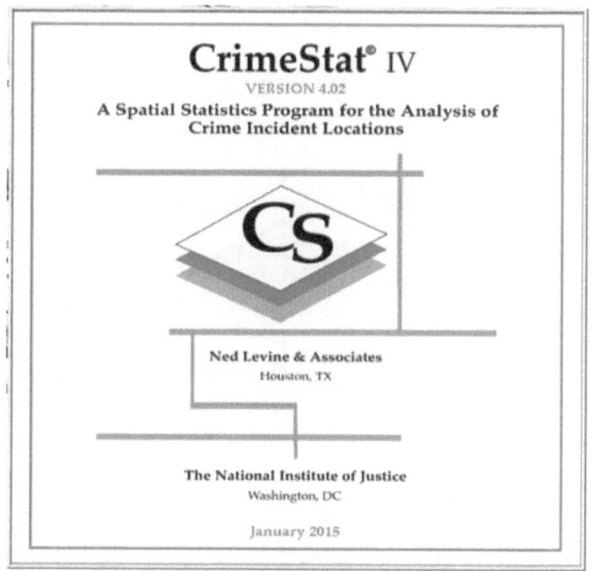

Figure 2.3 Screenshot representation of the CrimeStat software.

objects to GIS files that can be imported into a GIS program such as shapefiles.

2.7 Research Methodology

The research process for identification of accident hotspots is depicted in Figure 2.4.

2.8 Case Study

The region of interest for this study is Telangana state, which encloses the city of Hyderabad. Road traffic accidents became a major problem in India, predominantly in Hyderabad. With the growing number of top education institutes, giant international companies, and IT-based firms opt for Hyderabad as a trade center. A lot of job opportunities in the city have opened up for both technical and non-technical positions and too many people migrating over there end up making Hyderabad a commercial landmark, which is ironic. Due to urbanization, citizens are facing problems in day-to-day life like traffic congestion, increasing vehicular traffic, pollution, and could even worsen if

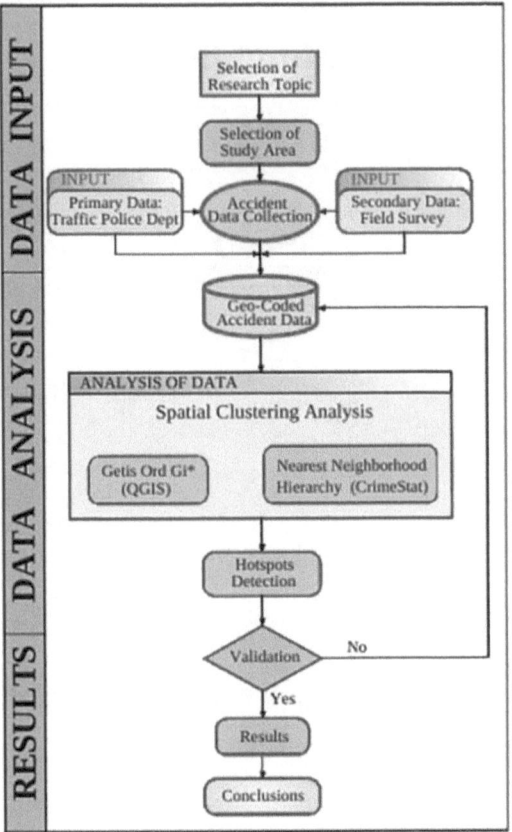

Figure 2.4 Research process.

the infrastructure is not upgraded to meet population demands. The city has a dense road network with a greater number of road traffic accident cases, making it an ideal location for the intended study as shown in Figure 2.5.

2.9 Analysis

The research process presented in Figure 2.4 is demonstrated through a case study shown in Figure 2.5 of a road network of the city of Hyderabad in the State of Telangana using the application of Getis-Ord Gi* and NNH methods for the identification of traffic accident hotspots. A stepwise procedure for the identification of traffic accident hotspots is presented below:

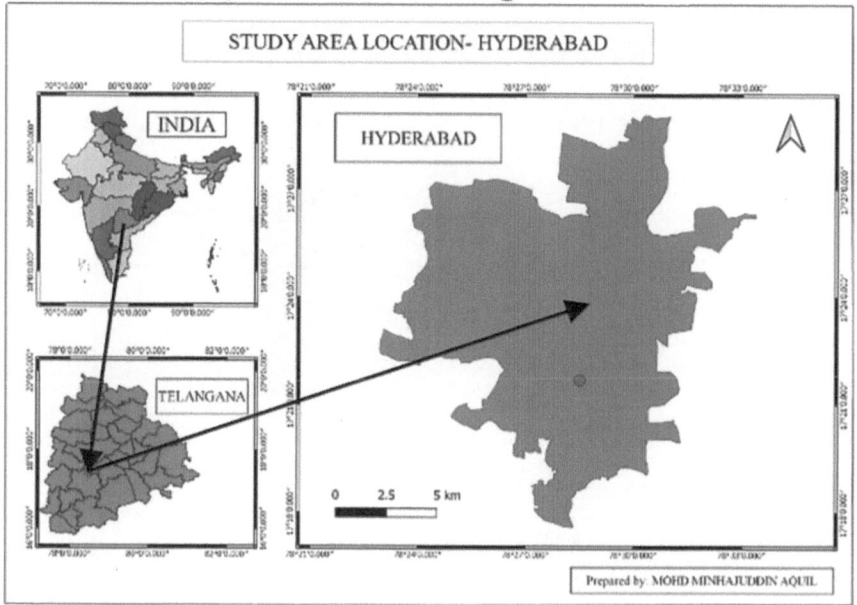

Figure 2.5   Study area location.

Step 1: Data Input in QGIS

Five years of accident data are collected from the Hyderabad traffic police department for research work. Geocoding is done and a .csv file is added in vector data in QGIS as shown in Figure 2.6.

Step 2: Formation of Grids in QGIS

The data are represented graphically in the form of a grid. Getis-Ord Gi* can be performed after developing a uniform pattern of grids for analysis of the study area as shown in Figure 2.7.

Step 3: Identification of Significant Hotspots

Hotspot analysis is performed using Getis-Ord Gi*. The output represents strategically significant hotspots and cold spot values as shown in Figure 2.8.

Step 4: Hotspot Data input

Accident hotspot data are uploaded in CrimeStat as shown in Figure 2.9.

Step 5: Analysis of Hotspots—NNH

Accident hotspot data analysis is performed using NNH in CrimeStat and the output file is saved as .shp as shown in Figure 2.10.

Step 6: Visualization of NNH Clustering—QGIS

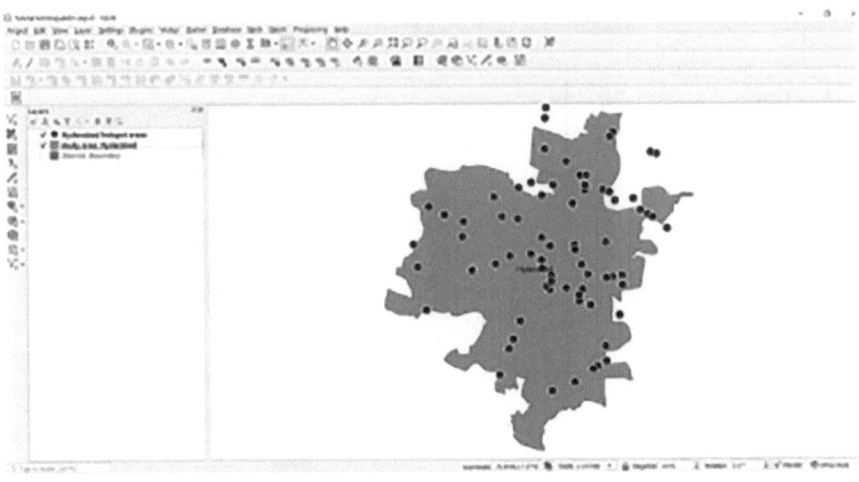

Figure 2.6　Showing hotspot areas in Hyderabad.

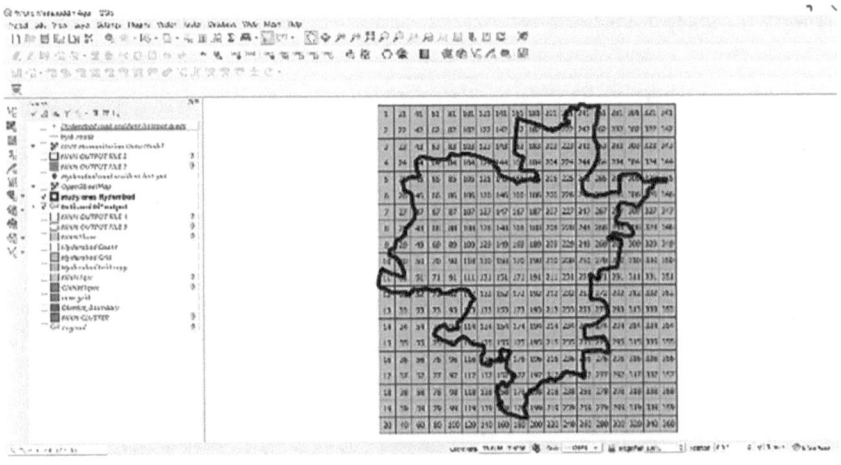

Figure 2.7　Formation of a grid in the study area limits.

The output of the CrimeStat file is uploaded in QGIS for visualizing the clustering patterns, which are shown in the form of first- and second-order ellipses and convex hulls. First-order clustering defines the center of minimum distance as a cluster center as shown in Figure 2.11. Convex hulls draw a polygon around the points in the cluster pattern as shown in Figure 2.12.

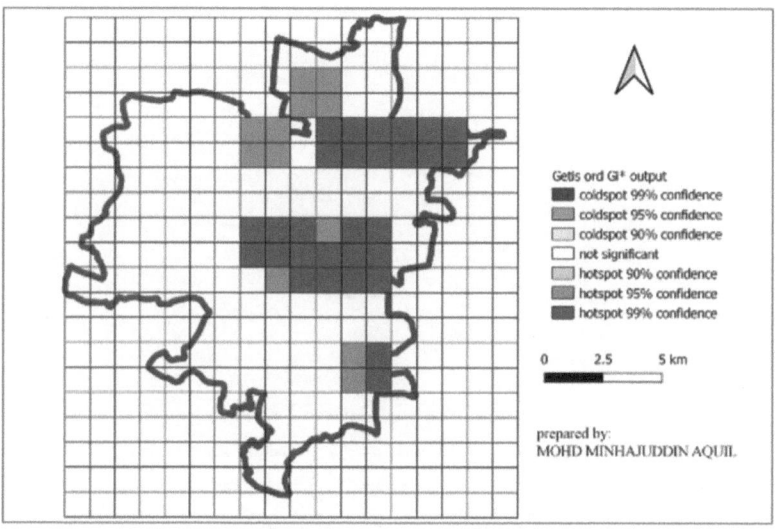

Figure 2.8   Strategically significant hotspots.

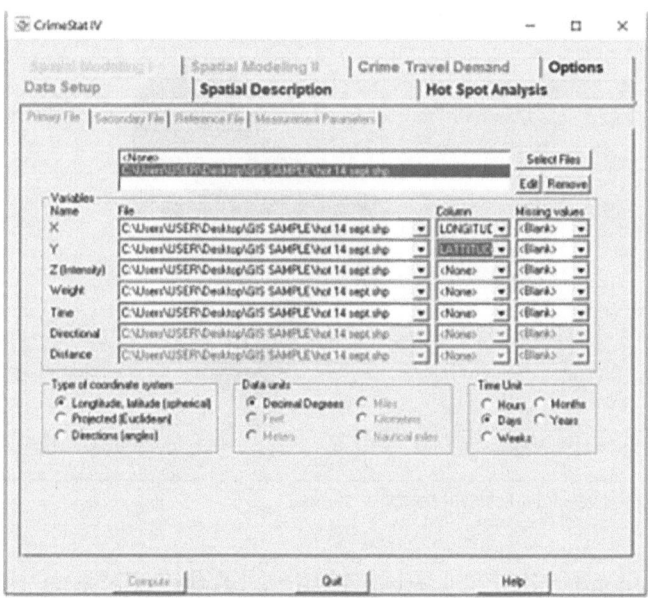

Figure 2.9   Data input in the CrimeStat software.

Substantially, first-order clusters are grouped into second-order clusters as shown in Figure 2.13.

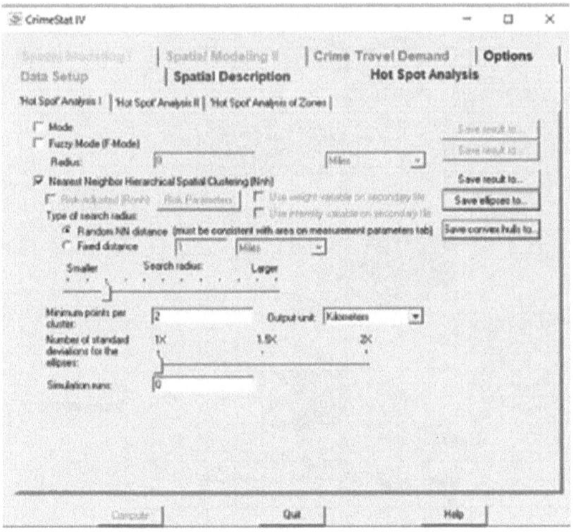

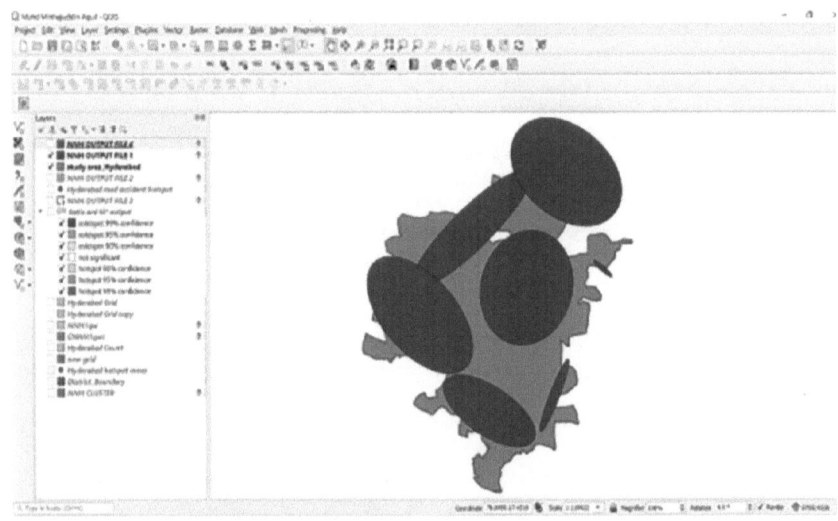

Figure 2.10   Hotspot analysis using NNH in the CrimeStat software.

Figure 2.11   First-order accident hotspots: ellipses.

The second-order clustering pattern of the ellipse is shown in Figure 2.14.

Step 7: **Visualization of overlapped Ellipses and Convex Hulls in QGIS**

The first- and second-order ellipses are overlapped and visualized in QGIS.

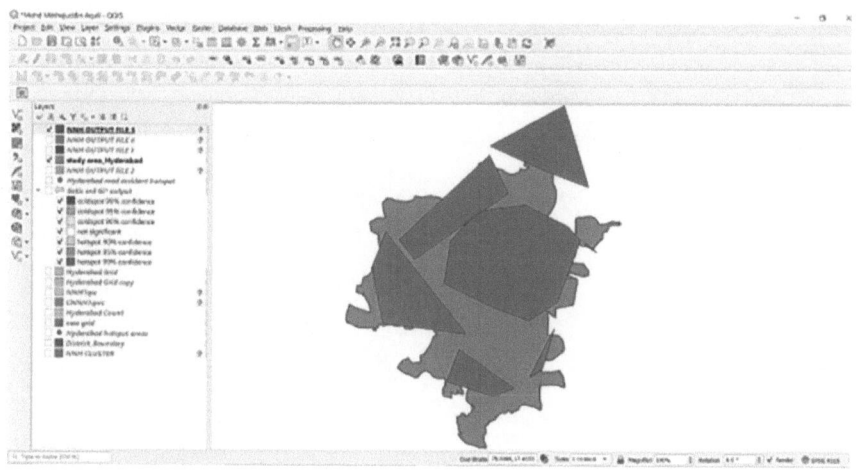

Figure 2.12 First-order hotspots: convex hulls.

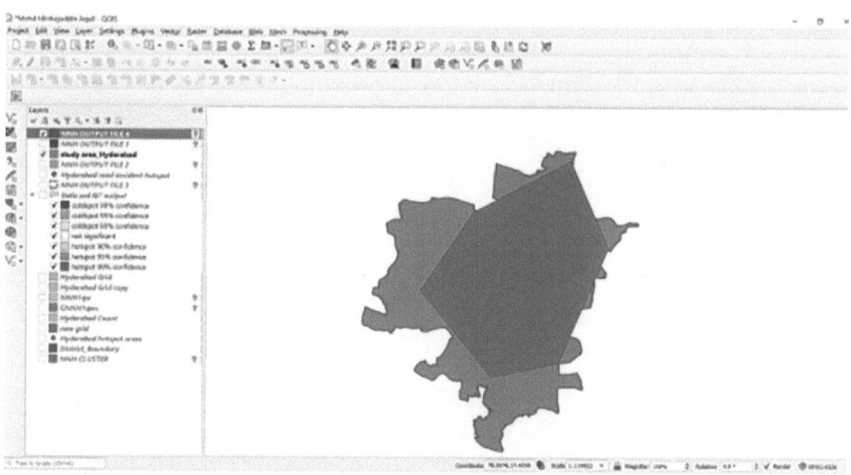

Figure 2.13 Second-order accident hotspots: convex hull.

It can be clearly observed that the regions having a greater number of accident hotspot locations are clustered as shown in Figure 2.15. The NNH routine has six outputs such as order of cluster pattern, number of points, number of clusters, standard deviation of ellipse, area of an ellipse, and density of cluster. First- and second-order convex hulls are overlapped and visualized in QGIS as shown in Figure 2.16.

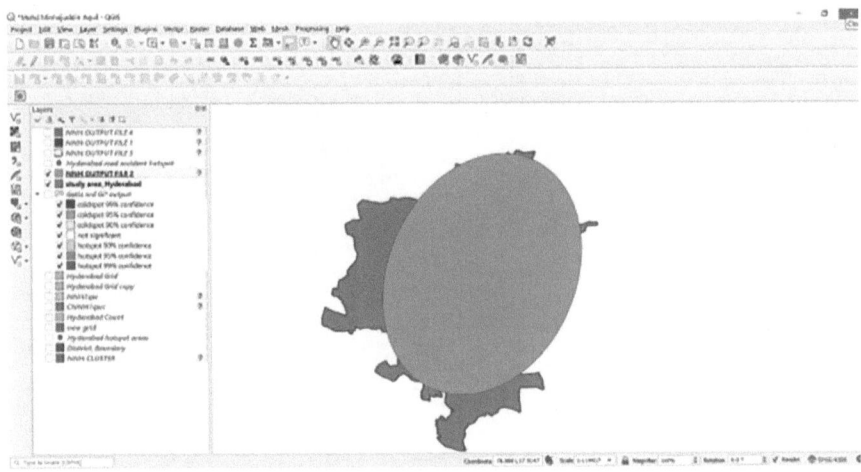

Figure 2.14   Second-order accident hotspots: an ellipse.

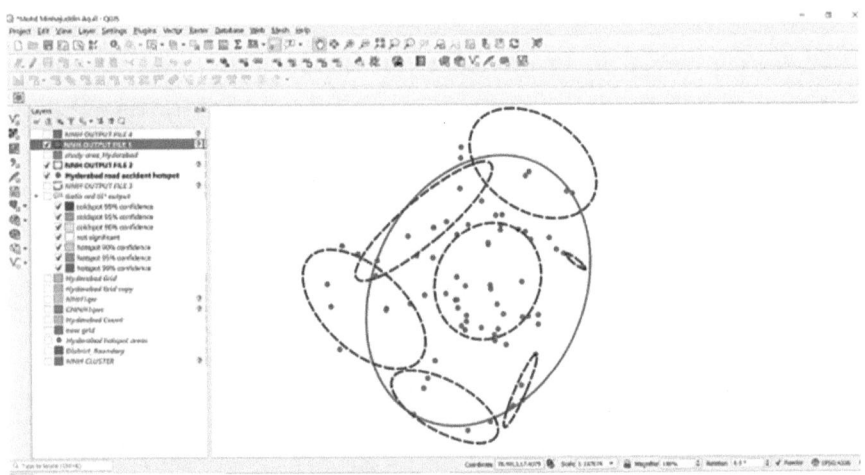

Figure 2.15   First- and second-order accident hotspots: ellipses.

Step 8: Combined Clustering Pattern of Hotspots

Overlapping of two methods of hotspot analysis is shown in Figure 2.17.

2.10 Results and Discussions

The NNH and Getis-Ord Gi* geospatial models were developed and employed for analyzing traffic accident hotspots considering the

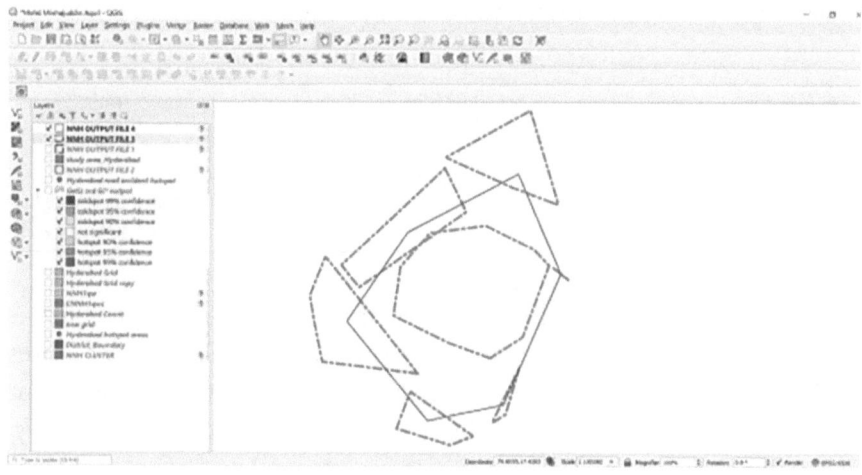

Figure 2.16 First- and second-order hotspots: convex hulls.

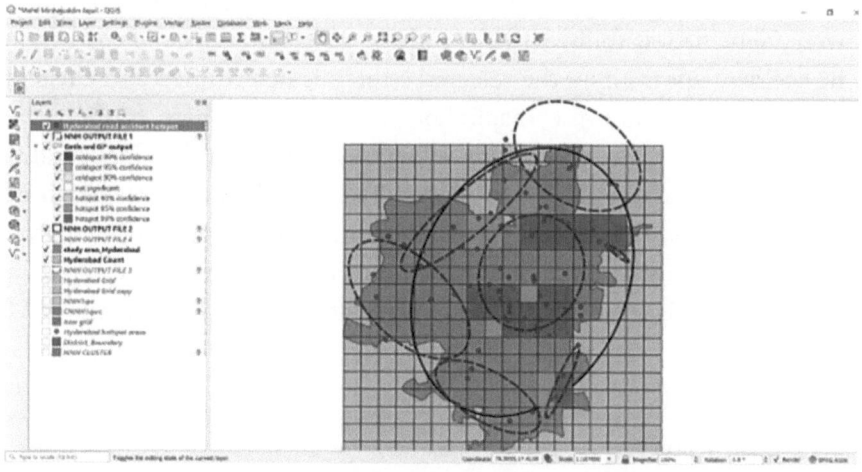

Figure 2.17 Overlapping of Getis-Ord Gi* and NNH methods.

road network of the city of Hyderabad. The results obtained using the Getis-Ord Gi* model resulted in high and low clustering values. This classification provides a color-coded map that gives a clear visualization of road traffic accident hotspots. An increase in the degree of redness indicates the higher risk of accident-prone regions as shown in Figure 2.8. NNH shows clusters of the first and second orders as presented in Figures 2.15 and 2.16. In NNH clustering, it is observed that convex hulls cover exactly all clustered points of road

Table 2.1 Traffic Accident Hotspot Locations

	SPATIAL CLUSTERING ANALYSIS	
PARAMETER	GETIS-ORD GI*	NNH
High- and Low-Value Clusters	Identified as referred in Figures 2.8 and 2.12	—
Convex Hulls and Polygons	—	Identified as referred in Figures 2.15 and 2.16

traffic accident hotspots. Furthermore, the boundary of convex hulls is specified from the points identified in the hotspot as shown in Figures 2.12, 2.13, and 2.16. On the other hand, ellipses not only represent the hotspots but also beyond them as shown in Figure 2.15. Finally, Getis-Ord Gi* and NNH outputs are overlapped and an overall clustering was observed in the spatial environment as shown in Figure 2.17. Both methods depict a similar clustering pattern and according to the study, the areas under high risk of road traffic accident hotspots were identified as follows: the top three 99% confidence of road traffic accident spots along the selected stretch is determined and clusters are formed to show the major hotspot locations where traffic accidents are occurring, Nampally, Malakpet, Masabtank, Chadergaht, Barkatpura, Santoshnagar, Secunderabad, and Tarnaka, as shown in Table 2.1. Research questions arise such as which method of spatial clustering needs to be adopted for more accuracy which resulted in further investigation of the high necessity to assert using the PAI Index.

2.11 Conclusion

This research explored the qualitative and quantitative traffic parameters influencing the accident hotspots on hazardous locations. The GIS method is employed to evaluate vehicular road accidents in complex networks. This research can be concluded as follows: (1) with access to data on accidents and traffic, we can calculate the risk measures, which will help us to identify safety levels on selected elements of the road. (2) Heterogeneity of traffic is another problem that causes severe congestion. This research grants an insight into the present scenario of the traffic condition and the most road traffic accident-prone locations in the study area. (3) The signals should be monitored at nighttime to control the night traffic. Mainly in the nighttime, nobody follows the signal, which is another cause of accidents. (4) Potholes are another cause of accidents because these are not visible clearly during the nighttime for two-wheelers. Other spatial analysis methods can be employed for further research like IDW, Kriging, KDE, and so on. It is recommended to consider the total number of injuries and fatalities for reducing urban road traffic accident hotspots.

References

Faheem M.I. et al., "Identification of blackspots and development of prediction model using multiple linear regression model", *International Journal of Research and Analytical Reviews*, Vol. 6, No.2, pp. 680–687, 2019.

Faheem M.I. et al., "A GIS approach road infrastructure safety management using GIS", *International Journal of Technical Innovation in Modern Engineering and Science*, Vol. 4, No. 8, pp. 250–261, 2018.

S.Sarifah. et al., "Determining hotspots of road accidents using spatial analysis", *Indonesian Journal of Electrical Engineering and Computer Science*, Vol. 9, No.1, pp.146–151, 2018.

Gebretensay B.F. et al., "Road traffic accident analysis and prediction model: A case study of Vadodara city", *International Research Journal of Engineering and Technology*, Vol. 5, No. 1, pp. 191–196, 2018.

Faheem M.I. et al., "A GIS approach for identification of accident hotspots and improvement at intersections in Hyderabad city", *International Journal of Engineering Research and Industrial Applications*, Vol. 10 No. 2, pp. 13–23, 2017.

Pandey et al., "Identification and improvement of accident black spots on N.H.86 district Sagar, Madhya Pradesh", *International Research Journal of Engineering and Technology*, Vol. 04, No. 09, pp. 740–742, 2017.

Anitha S et al., "Identification of hotspots of traffic accidents using GIS", *International Journal of Advances in Engineering and Technology*, Vol. 7, No. 3, pp. 13–16, 2016.

Thakali et al., "Identification of crash hotspots using kernel density estimation and kriging methods: a comparison", *Journal of Modern Transportation*. 2014.

Bhagyaiah M. et al., "Traffic analysis and road accidents: A case study of Hyderabad using GIS", *7th IGRSM International Remote Sensing & GIS Conference and Exhibition*, pp. 1–9, 2014.

Carola A. Blazquez et al., "A spatial and temporal analysis of child pedestrian crashes in Santiago, Chile", *Journal of Accident Analysis and Prevention*, pp. 304–311, 2013.

Prasannakumar et al., "Spatio-temporal clustering of road accidents: GIS based analysis and assessment", *International Conference: Spatial Thinking and Geographic Information Sciences*, pp. 317–325, 2011.

Anderson T.K., "Kernel density estimation and K-means clustering to profile road accident hotspots", *Accident Analysis and Prevention*, Vol. 41, No. (3), pp. 359–364, 2009.

Erdogan et al., "Geographical information systems aided traffic accident analysis system case study: city of Afyonkarahisar", *Accident Analysis and Prevention*, pp. 174–181, 2008.

3

MACHINE-LEARNING ALGORITHMS ON COGNITIVE MULTISENSOR IMAGE APPLICATION

T. VENKATAKRISHNAMOORTHY

Sasi Institute of Technology and Engineering

M. DHARANI

Annamacharya Institute of Technology and Sciences

P. ANILKUMAR

CVR College of Engineering

Contents

3.1 Introduction about Machine Learning

Machine learning (ML) is the study of computer automatic data algorithms for data analysis, which is based on sample data or training data to make predictions without being programmed. These predictions are based on computational statistics. Machining learning is the science of attainment of systems to act without being articulately

DOI: 10.1201/9781003102625-3

programmed. It has an important role in reducing the complexity of data learning systems. In previous days, it was very difficult to handle large data analysis with human intervention, requiring commercial software or high human intervention; also sophisticated results were not obtained in the case of large data analysis. ML is one of the best methods for automatically analyzing and studying data; it is a part of Artificial intelligence (AI) to get an idea of learning from data with the help of a system for making decisions and identifying the patterns. Through this procedure, we can construct computer programs, develop suitable solutions, and improve with experience [1]. In the past decade, ML algorithms are limited to some fixed applications such as controlling a car, web search, and controlling the understanding of the human genome. Nowadays, ML algorithms are pervasive in various fields to progress toward human-level AI.

Image processing is one of the demands in the industry that seems to improve every year. It is applied for a complex algorithm to input images for technical analysis and returns useful information as output. In this process, the system making the automatic analysis performs to deliver efficient results at a fast rate. In advanced techniques, many libraries and frameworks have been implanted as a tool for solving image-processing problems with ML techniques. The data analysis algorithms are widely implemented in various image-processing techniques based on applications. The detection of any object in local or global analysis data is difficult in a large dataset by using standard image processing algorithms. This problem can be overcome by implementing advanced data analysis algorithms. In real-time applications, the image processing analysis can be performed in various fields such as medical, commercial, satellite image and remote sensing, and so on. The ML algorithms can be categorized into unsupervised, supervised, and reinforcement learning methods to perform the data analysis, classification, and prediction of the information. Training sets are needed for performing the supervised classification, and the computer can select the classes in unsupervised classification. In the reinforcement learning method, a computer program interacts with a dynamic environment to perform a certain goal.

3.2 Introduction to Remote Sensing and Satellite Image Processing

Satellite remote sensing is an art of science, which acquires or captures the information of Earth's surface without any physical contact and thus in contrast to on-site observation within less time. In simple terms, remote sensing acquires the Earth's surface information within the particular swath area for analyzing the earth science disciplines such as meteorology, oceanography, land surveying, geology, and military applications. In this technique, the data collected from the inaccessible area include monitoring the deforestation in areas, coastal and ocean depths, and the Arctic and Antarctic regions. It is very difficult to survey in these typical areas manually; therefore, data are collected through satellite remote sensing [2].

Satellite image processing is one of the important areas in research to analyze and monitor the Earth's information, which is collected from artificial satellite receivers. Initially, the Earth's information is captured by satellite sensors and later processed by computers to extract the information. Satellite imagery is widely implemented to plan the infrastructures or environmental conditions or to predict upcoming disasters. The processing of pixels is necessary to obtain an efficient analysis of the Earth's surface objects. It is one of the techniques of remote sensing, which is based on pixel resolution to collect coherent information about the Earth's surface objects [3].

3.3 Multispectral/Hyperspectral Satellite Images

The satellite sensors capture the Earth's information with various sensors and give the total collected information in the form of multispectral or hyperspectral images, which collect the information across the electromagnetic spectrum. The main function of this imaging is to perform the pixel operation in the image of the scene for identifying or classifying the object or detecting the process. The various types of scanners have captured the information for different resolutions. There is push broom, whisk broom, and band sequential scanners for the spatial and spectral scanning. The scanning process can be performed to get the acquired images of an area with different frequencies, which give different information based on the applications. Generally, there are four types of resolutions available for multispectral or hyperspectral resolutions such as spatial, spectral, temporal, and radiometric resolutions [4,5].

3.4 Dimensional Reduction Technique

ML algorithms are one of the good sources in the case of dimensional reduction techniques, which are helpful in many classification processes. The classification is based on too many factors also called as variables. More features are getting to visualize the training set for implementing the work. In multispectral or hyperspectral satellite images, more features are correlated, so automatically more redundant data are available in the original input image [6,7]. Data size and complex image processing are needed to avoid these problems. Instead of these, the dimensionality reduction algorithms come into play to reduce the redundant data by reducing the number of random variables or factors. It is the mapping process that converts from a higher dimensional to a lower dimensional space by reducing the repeated information from the large dataset to good approximation to the efficient result [8].

There are linear and non-linear methods that are used in DR methods. The Principal Component Analysis (PCA) is the best method in linear dimensionality reduction. Compared with non-linear dimensional methods, PCA is suitable for removing the redundant information, for comparing, and for feature extraction of the images [9]. The Independent Component Analysis (ICA) is an example of the non-linear dimensionality reduction methods and works with higher-order statistics such as skewness, kurtosis, and negentropy. The ICA and PCA methods are used in many applications for reducing the complexity of large data analysis [10].

Figure 3.1 shows the multispectral image, which is collected from the NOAA satellite. This image has ten spectral bands. Figure 3.2 shows the SAR images, which contain more than hundred bands. These images contain a large number of bands, and it is difficult to analyze or process individual bands. The user also does not know the exact information of the bands. There are various standard image processing techniques such as filtering, enhancement, compression, and

Figure 3.1 Multispectral satellite image.

classification algorithms that are implemented to extract the feature of the local or global images with panchromatic (grayscale) and color images (RGB) [11,12]. In this process, there is a limited number of bands, which makes it easy to process all the bands in less time. This same process does not help in processing the large dataset of multi-spectral or hyperspectral data. Commercial software or simulators are required for processing the whole data. The study or processing of the whole bands is also not possible within less time. These limitations are overcome with dimensionality reduction techniques.

3.5 Results and Discussion

PCA/ICA is an unsupervised, non-parametric statistical technique primarily used for dimensionality reduction in ML [13,14]. Figure 3.3 shows the multispectral image, which is downloaded from NOAA-HRPT. The multispectral image contains ten bands and gives raw image information and spectral information. The image processing techniques are difficult to apply to each and every band information within a short duration. Instead of the image processing techniques, the dimensional reduction techniques are preferred to obtain the total number of band information into three numbers as shown in Figure 3.4

Each multi/hyperspectral image gives the object information in the form of spatial and spectral values. In this pre-processing technique,

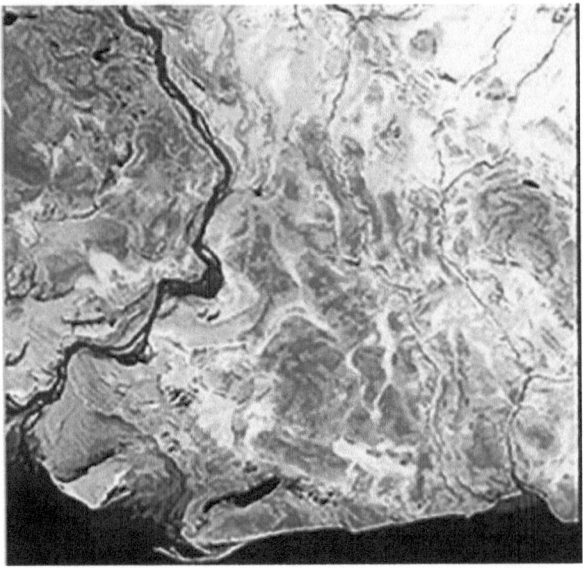

Figure 3.2   Hyperspectral image.

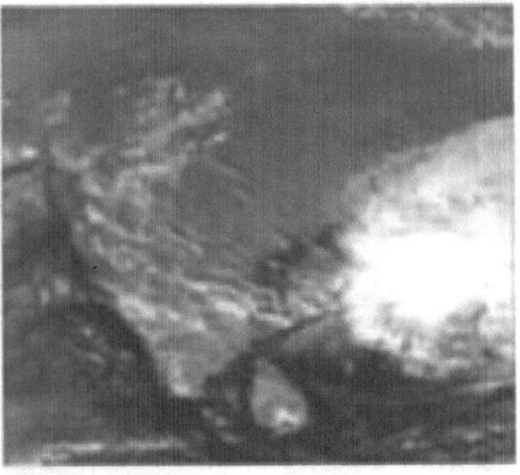

Figure 3.3   NOAA-multispectral image.

these values are lost, so automatically, the user gets misclassification results due to a lack of knowledge on the particular object on the local or global image. These image spatial and spectral values are retrieved by an enhancement technique.

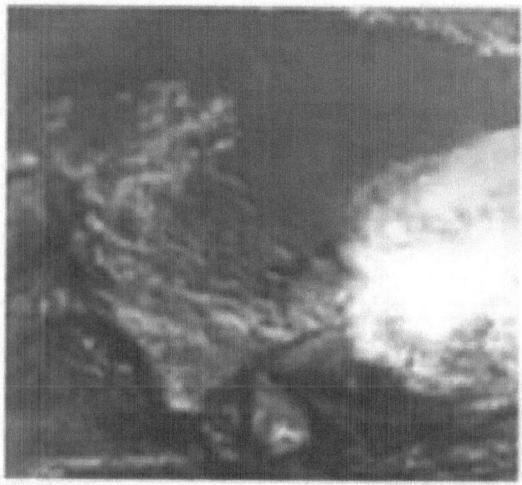

Figure 3.4 Proposed image.

Steps for the Proposed Algorithm:

1. Collect the multispectral satellite image from the NOAA-HRPT Receiver.
2. Perform the geometric and radiometric corrections for avoiding the avoid mismatches.
3. Perform the dimensional reduction technique by PCA.
4. Apply morphological operations to separate the lower and higher frequency details.
5. Apply HIS transform and histogram equalization to obtain the normalization of the color and brightness values.
6. Apply the inverse PCA technique for retrieving the original image.

Table 3.1 shows the comparison table. HIS transformation is one of the best techniques for improvement. Brovey transforms are limited to three-band multispectral images only. The PCA and ICA ML algorithms are suitable for dimensional reduction techniques, which are more beneficial than the standard compression technique [8,15]. But they do not maintain tradeoff levels between spatial and spectral values. For obtaining the loss components, implement the morphological and wavelet transformation and normalize these values with the histogram matching technique. So, the color, bright, low, and

Table 3.1   Comparison of Results with Different Methods and the Proposed Method

IMAGE 1	PSNR	ERGAS	RASE	ENTROPY
HIS transformation	33.24	14.94	49.31	6.71
Brovey transformation	28.38	18.23	51.25	7.51
Principal component analysis	22.13	20.56	48.24	6.97
Independent component analysis	28.32	14.58	47.12	7.47
Proposed method	**36.24**	**11.79**	**44.21**	**7.57**

high-frequency components are retrieved by using this proposed technique, and values are measured with the Peak Signal-to-Noise Ratio (PSNR), Relative Global Dimensional Synthesis Error (ERGAS), and entropy values. The PSNR values are high and ERGAS is low for enhanced images with good histogram values.

3.6 Conclusion and Future Scope

The multispectral/hyperspectral satellite images have a high number of bands with low resolution. The user has not identified or detected the object, so automatically, there is misclassification of results due to the lack of object information. Direct images are difficult to process and detect objects. The ML algorithms are applied to reduce the redundant information between the bands and improve the spatial and spectral values. These algorithms are very important in remote sensing applications such as land surveying, studying vegetation levels, and predicting atmospheric levels. Including these applications, the ML algorithms are very applicable to medical images also.

References

1. Chen S., Wang H., Xu F., and Jin Y.Q., "Target classification using the deep convolutional networks for SAR images," *IEEE Transactions on Geoscience and Remote Sensing*, 2016, Vol. 54, PP: 4806 –4817.
2. Pohl C. and van Genderen J.L., "Multisensor image fusion in remote sensing: Concepts, methods and applications," *International Journal of Remote Sensing*, 1998, Vol. 19, pp. 823–854.
3. Anuradha B. and Anil Kumar P. "Determination of convective cloud parameters using WRF and DWR data," *Journal of Advanced Research in Dynamical and Control Systems*, 2016, Vol. 4, No. 3, pp. 1715–1724.

4. Hryvachevskyi A., Prudyus I., Lazko L., and Fabirovskyy S., "Improvement of segmentation quality of multispectral images by increasing resolution," in 2nd International Conference on Information and Telecommunication Technologies and Radio Electronics, UkrMiCo 2017- Proceedings 8095371.

5. Chibani Y. and Houacine A., "The joint use of IHS transform and redundant wavelet decomposition for fusing multispectral and panchromatic images," *International Journal of Remote Sensing*, 2002, Vol. 23, No. 18, pp. 3821–3833.

6. Shahid Z., Dupont F., and Baskurt A., "A novel efficient image compression system based on independent component analysis," *Electronic Imaging, SPIE*, 2009, Vol. 7248, pp. 724808-1–724808-9.

7. Venkatakrishnamoorthy T. and Umamaheswara Reddy G., "Image classification using higher-order statistics-based ICA for NOAA multispectral satellite image," *International Journal of Advanced Intelligence Paradigms*, Inder science publishers. 2020. Vol. 17, No. 1/2, pp. 178–191.

8. Devi K.I. and Shanmugalakshmi R., "An efficient image compression technique using 2-dimensional principal component analysis," *European Journal of Scientific Research*, ISSN 1450-216X, 2011, Vol. 58, No.4, pp. 497–505.

9. Jackson J.E. *A User's Guide to Principal Components*. Vol. 587 John Wiley & Sons, (Newyork, Singapore); 1991.

10. Wichern D.W. and Johnson R.A., *Applied Multivariate Statistical Analysis*. 6th ed. Pearson Education Limited; 2014.

11. Sangwine S.J. and Horne R.E.N., *The Colour Image Processing Handbook*. Chapman & Hall; 1989.

12. Gonzales R. and Woods R., *Digital Image Processing*. 3rd ed. Prentice-Hall; 2008. ISBN 9780131687288.

13. Dharani M. and Sreenivasulu G., "Land classification of land sat multispectral image using principal component analysis and morphological operations," *Journal of Advanced Research in Dynamical and Control Systems (JARDCS)*, 2018, Vol. 10, No. 6, pp. 164–173.

14. Bhaskarrajan N.J., "Satellite image fusion using IHS and PCA method," *International Journal of Innovative Science Engineering and Technology*, 2014, Vol. 1, pp. 152–156.

15. K.I. Devi, R. Shanmugalakshmi, "An efficient image compression technique using 2-dimensional principal component analysis," *European Journal of Scientific Research*, 2011, Vol. 58, No. 4, pp. 497–505.

4

Performance Evaluation of Video Compression Techniques: x.263, x.264 and x.265 to Improve Video Streaming Quality

B. SHILPA

KLEF University

B. ANIL KUMAR

GRIET University

L. KOTESWARARAO

KLEF University

Contents

DOI: 10.1201/9781003102625-4

4.1 Introduction

Vehicular communication is the most evolving application where passengers rely on entertainment and pass their time on video streaming. Apart from that, passengers are dependent on wireless communications for other applications like safety, traffic control, live streaming. It is a challenging task to play a video without discontinuity as the user travels from one cell zone to another zone. Provision of such high-quality video streaming is a challenging task in wireless networks. Wide bandwidth for vehicular networks with low latency and high capacity even for the uncompressed data (Figure 4.1) is provided by the 5G network [1–5]. But compressing is essential for any application where we can reduce the bandwidth and storage capacity. Therefore, we choose different compressing algorithms to satisfy the requirements. In recent years, there is a tremendous improvement in video coding techniques, which play a major role in many applications such as video conferencing, streaming, live streaming, video on demand and ultra HD TV [6]. These occupy most of the storage capacity and need to be converted before transmission. The process of changing the size or reducing the bit rate is called compression. In this process, the bit size is reduced by choosing only the important elements and discarding some redundant bits.

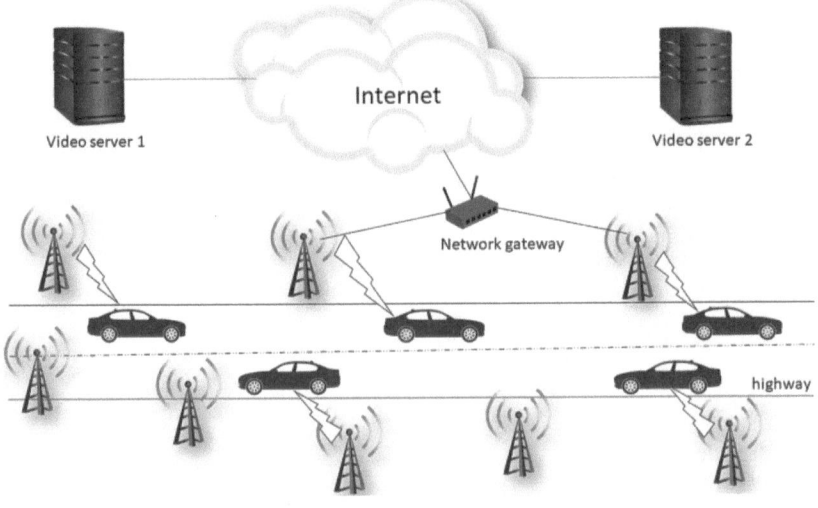

Figure 4.1   Video streaming in 5G enabled vehicular networks.

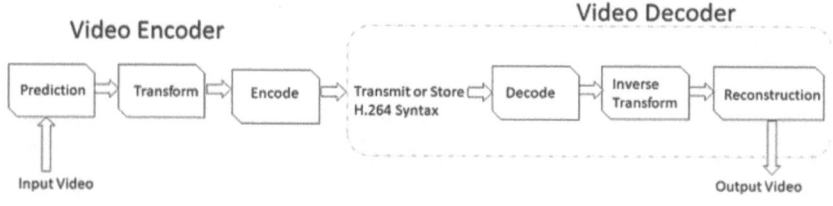

Figure 4.2 Basic encoder and decoder.

The method of video compression involves both encoding and decoding as shown in Figure 4.2.

The Encoder transforms the raw digital video to the compressed form, which can be stored and transmitted.

The Decoder transforms the compressed video back to the original video.

We can classify the compression methods into lossless video compression and lossy video compression. In the lossless video compression, the output exactly matches the original input, whereas in the lossy technique, the output has missed details while performing the compression. We have various algorithms for compression. But we limit our discussion to terms that were formulated by international standards [7--8]. The reason for choosing the standards is the feature of interoperability, which gives the flexibility to decode the video data anywhere. The compression technique for still images is described by the Joint Photographic Experts Group (JPEG). We can have a brief discussion on MPEG (Moving Pictures Expert Group) standards for video compression, which is a group formed by the International Organization for Standardization (ISO) and the International Electrotechnical Commission (IEC) [9]. MPEG was given a formal status within the ISO/IEC. The original work started by the MPEG group culminated in the standard called MPEG-1 in 1992, ISO/IEC 11172 [1]. The MPEG-1 standard itself comprises five parts. They are part 1: systems; part 2: video; part 3: audio; part 4: compliance testing; and part 5: software simulation. MPEG-1 has been targeted for multimedia applications. It was optimized for compressing progressively scanned CIF (Common Intermediate Format) images of size 352×240 pixels at 30 frames per second (fps) at data rates of about 1.5 Mb/s.

Table 4.1   Different Frame Video Resolutions

FRAME FORMAT	RESOLUTION (HORZ. × VERT.)	BITS PER SAMPLE
Sub-QCIF	128 × 96	147,456
Quarter CI (QCIF)	176 × 44	304,128
CIF	352 × 288	1,216,512
4CIF	704 × 576	4,866,048

Table 4.2   Different Frame Formats

RESOLUTION	MEASUREMENTS (IN PIXELS)	PIXEL COUNT
4k (UHD)	3,840 × 2,160	48,294,400
1080p (Full HD)	1,920 × 1,080	2,073,600
720p (HD Ready)	1,280 × 720	921,600
480 p (SD)	640 × 480	307,000

4.1.1 Video Formats

There are many different formats in which a video is presented. We can transform the captured into any one form of an intermediate format. CIF is the simple and most used format [10]. The figure 4.2 shows the example of video format sampled at 4CIF, CIF, QCIF and SQCIF as shown in Table 4.1. The selection of a frame depends on many factors such as storage and number of bits transmitted. Sub-quarter CIF is the smallest standard image size. With its resolution of 128×96 pixels, it provides low-resolution video clips and streaming video on mobile phones. QCIF resolution is four times smaller than the CIF resolution. QCIF is used for multimedia applications (Table 4.2).

4.2 Selected Codecs

4.2.1 H.264

It is the most popular document published by international standard bodies ITU-T (International Telecommunication Union) and ISO/IEC (International Organization for Standardization/International Electrotechnical Commission). It was proposed in 2003 by the Joint Video Team (JVT) that consisted of hundreds of video compression experts from the MPEG and the Video Coding Experts Group (VCEG) [11].

The main advantage of H.264 compared to previous standards is that it has four times better quality at the same compressed rate as MPEG-4 and has data rates that are lower than MPEG-2. There is a tradeoff between better compression performance of H.264 and computation cost and it also consumes more power for compressing and decompressing [12–14]. This standard supports packet format also to reduce transmission errors. Other applications include HD-DVD and Blu-Ray formats and Internet video videoconferencing (Figures 4.3 and 4.4).

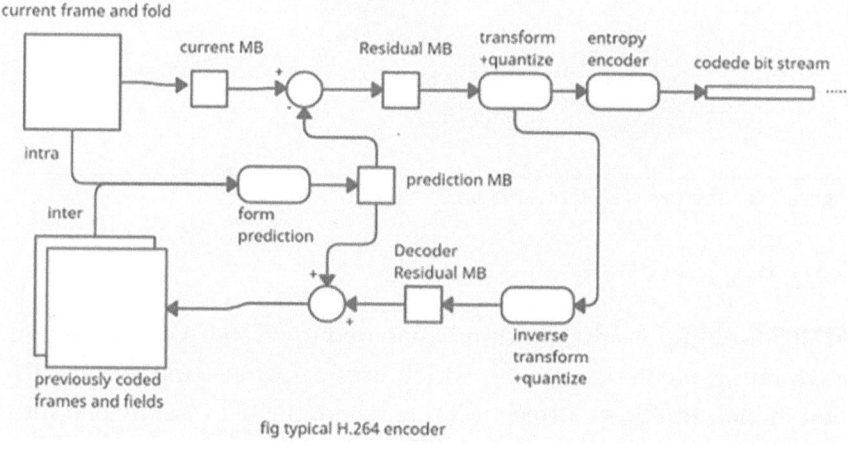

Figure 4.3 Typical H.264 encoder.

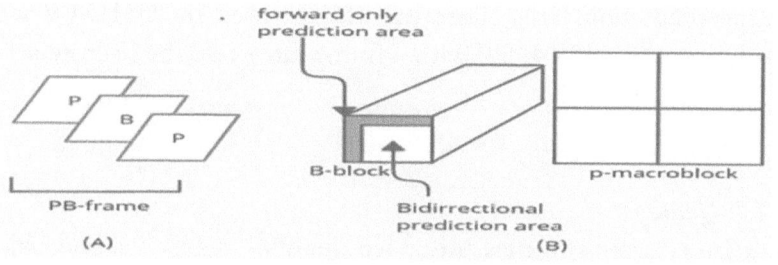

Figure 4.4 PB-frame prediction and B-block prediction in a PB-frame.

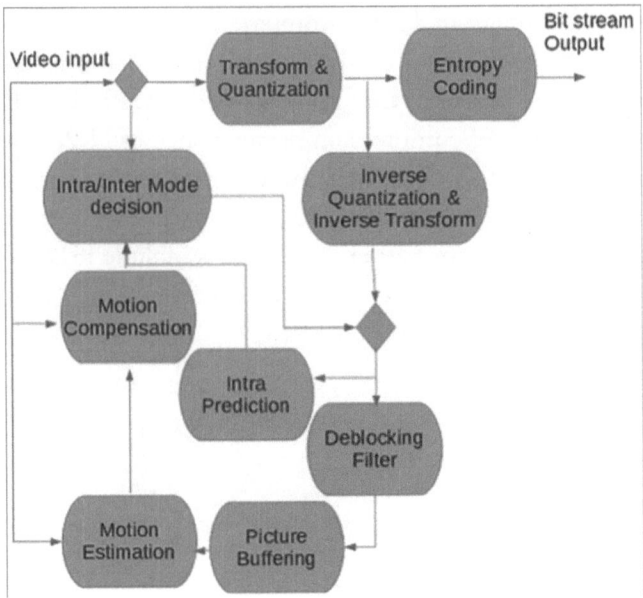

Figure 4.5   Overview of HEVC encoding and decoding.

4.2.2 H.265

H.265 uses improved algorithms compared to H.264/AVC. It uses an intracoding prediction model, which uses a spatial-temporal prediction model. It follows a tree structure where the CTU is divided into several blocks that are further divided into coding units where these units are used for prediction (intra or inter) [15]. The predicted code is transformed using transform coding and the difference of predicted and the current block is called residual, which is also used for performing prediction. This codec can also be used for UHDTV with a resolution of 8,192×4,320 with a frame rate of 30 fps (Figure 4.5).

4.2.3 Evaluation Metrics

4.2.3.1 Quality

There are two ways to measure video quality – subjective and objective. We cannot solely depend upon subjective measurement factors as they are not accurate and depend on the interaction between the components of the Human Visual system (HVS), the eye and the brain. Two factors are influenced: spatial fidelity (clear frame visibility) and temporal fidelity (with respect to time, how natural it is). There are a lot of factors such as the viewing environment, observer's state of mind and a non-disturbing and comfortable environment.

4.2.4 PSNR

PSNR is one of the evaluation metrics to evaluate compression [16]. The compression may be lossy or lossless. It is calculated as follows:

$$\text{MSE} = \frac{1}{MN} \sum_{J=1}^{N} \sum_{I=1}^{M} \left(X_{I,J} - Y_{I,J} \right)^2$$

$$\text{PSNR} = 10 \log \frac{\text{MAX}_I^2}{\text{MSE}}$$

It is a difficult task to measure the PSNR metric as it requires the use of an original image, which is not always possible in each case. The quality of the video is directly proportional to PSNR; low PSNR indicates low quality and high PSNR indicates high quality.

4.2.5 H.266

H.266 is the latest codec, which is developed by the Joint Video Experts Group (JVET) and the working group of MPEG, a group of ISO/IEC JTC. The other name for it is Versatile Video Coding (VVC). It is the descendant of H.265 and H.264, which is also known for High-Efficiency Video Coding (HEVC). Compared to AVC and HEVC, the performance of VVC is faster and it uses only half the data to that of HEVC. VVC uses the technology of UHD. The widest range of this usage is in mobile networks, particularly in video transmission [17–19]. As this technique uses ultrahigh-resolution streaming of 4k or 8k videos, it is beneficial and we can even use this for 360° video panorama, moving images, screen sharing and so on (Figures 4.6–4.8).

4.2.6 Evaluation Method

4.2.6.1 Results

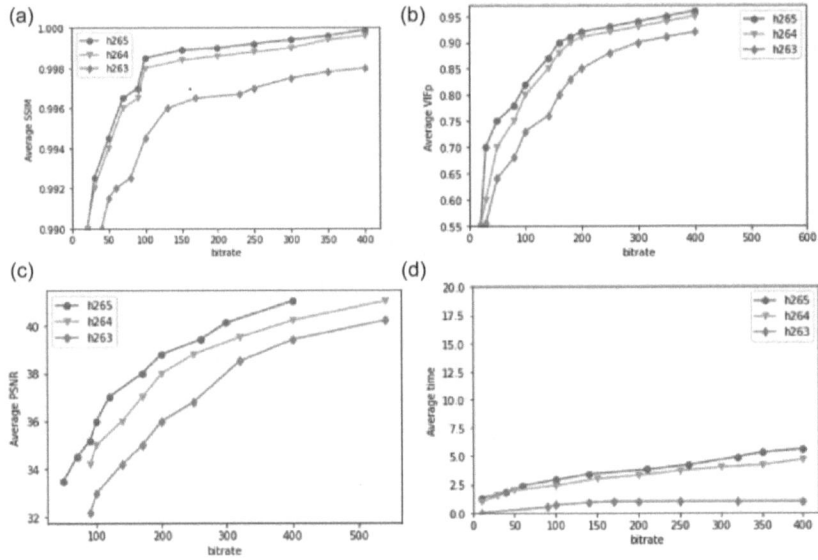

Figure 4.6 (a) SSIM with varying bit rates (file: Akiyo). (b) VIFp with varying bit rates (file: Akiyo). (c) Average PSNR for varying bit rates (file: Akiyo). (d) Average encoding time (file: Akiyo).

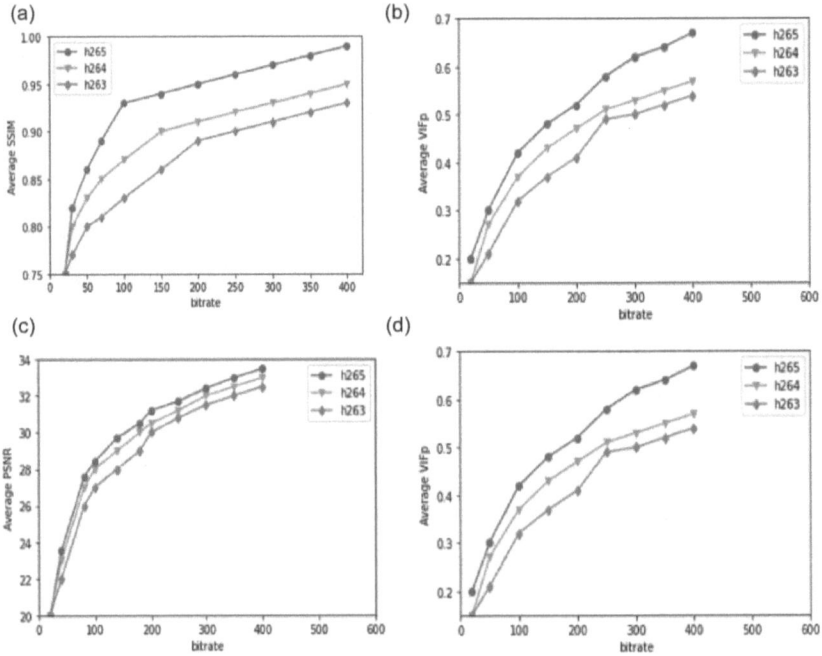

Figure 4.7 (a) SSIM with varying bit rates (file: coastguard). (b) VIFp with varying bit rates (file: coastguard). (c) Average PSNR for varying bit rates (file: coastguard). (d) Average encoding time (file: coastguard).

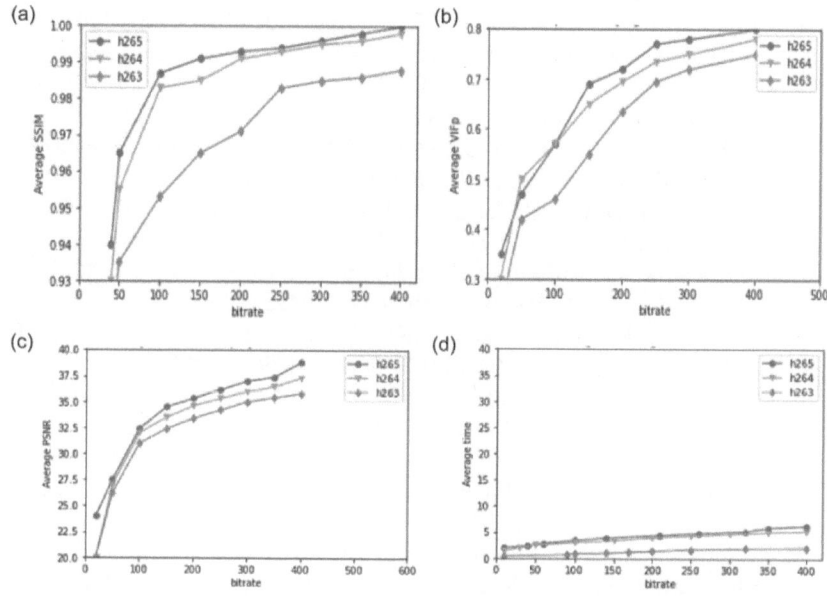

Figure 4.8 (a) SSIM with varying bit rates (file: Foreman). (b) VIFp with varying bit rates (file: Foreman). (c) Average PSNR for varying bit rates (file: Foreman). (d) Average encoding time (file: Foreman).

4.3 Conclusion

In this chapter, we compared the performance of three dominant video encoding techniques. One of them is H.263, which is the basic video codec used prior to the remaining two techniques. H.264 and H.265 techniques are matured enough. H.264 is better in terms of quality and bandwidth. H.265 uses coding Tree Units to process the information and requires less bandwidth. Until now, H.265 performs the best while H.264 is slightly behind. However, the performance of H.264 is very much convincing to justify the formation of AO Media, the alliance for developing a royalty-free codec for next-generation video encoding. The future codec H.266 is upcoming and can be studied for future scope.

References

1. Y. Kabalci, 5G Mobile Communication Systems: Fundamentals, Challenges, and Key Technologies. In: Kabalci E., Kabalci Y. (eds.), *Smart Grids and Their Communication Systems. Energy Systems in Electrical Engineering.* Springer, Singapore, 2019, doi:10.1007/978-981-13-1768-2_10.

2. J. Qiao, Y. He and X. S. Shen, "Improving Video Streaming Quality in 5G Enabled Vehicular Networks," *IEEE Wireless Communications*, vol. 25, no. 2, pp. 133–139, April 2018, doi:10.1109/MWC.2018.1700173.

3. *Book Vasudev Bhaskaran, Konstantinos Konstantinides "IMAGE AND VIDEO COMPRESSION STANDARDS Algorithms and Architectures"* Springer, Boston, MA, doi:10.1007/978-1-4757-2358-8.

4. M. A. Layek et al., "Performance analysis of H.264, H.265, VP9 and AV1 video encoders," *2017 19th Asia-Pacific Network Operations and Management Symposium (APNOMS)*, Seoul, Korea (South), 2017, pp. 322–325, doi: 10.1109/APNOMS.2017.8094162.

5. K. Sayood, *Video Compression*, Khalid Sayood, pp. 633–674, ISBN 9780124157965, doi:10.1016/B978-0-12-415796-5.00019-3.

6. H.263 Video Compression. In: Furht B. (eds) *Encyclopedia of Multimedia*. Springer, Boston, MA, doi:10.1007/978-0-387-78414-4_329.

7. A. Sallam, O. Faragallah and E. S. El-Rabaie, "Comparative study of video compression techniques," *Menoufia Journal of Electronic Engineering Research*, vol. 27, 2018. doi:10.21608/mjeer.2019.64366.

8. K. Rijkse, "H.263: Video coding for low-bit-rate communication," *IEEE Communications Magazine*, vol. 34, no. 12, pp. 42–45, Dec. 1996, doi:10.1109/35.556485.

9. T. Wiegand, G. J. Sullivan, G. Bjontegaard, and A. Luthra, "Overview of the H. 264/AVC video coding standard," *IEEE Transactions on Circuits and Systems for Video Technology*, vol. 13, no. 7, pp. 560576, 2003.

10. S. Ma, X. Zhang, C. Jia, Z. Zhao, S. Wang and S. Wang, "Image and Video Compression With Neural Networks: A Review," *IEEE Transactions on Circuits and Systems for Video Technology*, vol. 30, no. 6, pp. 1683–1698, June 2020, doi:10.1109/TCSVT.2019.2910119.

11. Vetrivel, S., K. Suba, and G. Athisha, "An overview of h. 26x series and its applications," *International Journal of Engineering Science and Technology*, vol. 2, no. 9, pp. 4622–4631, 2010.

12. Nidhi and N. Aggarwal, "A review on Video Quality Assessment," *2014 Recent Advances in Engineering and Computational Sciences (RAECS)*, Chandigarh, India, 2014, pp. 1–6, doi:10.1109/RAECS.2014.6799645.

13. J. Eze, S. Zhang, E. Liu and E. Eze, "Cognitive radio technology-assisted vehicular ad-hoc networks (VANETs): Current status, challenges, and research trends," *2017 23rd International Conference on Automation and Computing (ICAC)*, Huddersfield, UK, 2017, pp. 1–6, doi:10.23919/IConAC.2017.8082035.

14. P. Lambert, W. De Neve, Y. Dhondt, R. Van de Walle, "Flexible macroblock ordering in H.264/AVC," *Journal of Visual Communication and Image Representation*, vol. 17, no. 2, pp. 358–375, 2006, ISSN 1047–3203, https://doi.org/10.1016/j.jvcir.2005.05.008.

15. T. S. Rappaport et al., "Millimeter Wave Mobile Communications for 5G Cellular: It Will Work!" in IEEE Access, vol. 1, pp. 335–349, 2013, doi: 10.1109/ACCESS.2013.2260813.

16. Z. Ren, C. Ye, M. Liu and M. Liu, "A Fast Intra Prediction Algorithm for H.264," 2009 First International Workshop on Education Technology and Computer Science, Wuhan, China, 2009, pp. 772–775, doi: 10.1109/ETCS.2009.434.

17. G. Pastuszak and A. Abramowski, "Algorithm and Architecture Design of the H.265/HEVC Intra Encoder," *IEEE Transactions on Circuits and Systems for Video Technology*, vol. 26, no. 1, pp. 210–222, Jan. 2016, doi:10.1109/TCSVT.2015.2428571.

18. K. S. Thyagarajan, *A book on Still Image and Video Compression with MATLAB*, ISBN: 978-1-118–09776-2.

19. A book on High-Efficiency Video Coding (HEVC) Algorithms and Architectures.

5

Design of CMOS Circuits for Cognitive Radio Application with Power Analysis

DAYADI LAKSHMAIAH

Sri Indu Institute of Engineering and Technology

SHAIK FAIROOZ

Malla Reddy Engineering College (Autonomous)

J. B. V. SUBRAHMNYAM

GIET University

A. SINDHUJA AND R. YADAGIRI RAO

Sri Indu Institute of Engineering and Technology

Contents

DOI: 10.1201/9781003102625-5

5.1 Introduction

A mathematics circuit plays a critical role in the VLSI technology. Mathematics block is frequently the main power overwhelming parts within a structure because the switching action is rather high. Other calculation implementations can provide solutions for a decreased power utilization advantage to increase the presentation of the entire system. INTEL technology proposed that the number of transistors in square inches could be doubled every 18–24 months. They also observed that over a period, with the growth in technology, an increase in the number of transistors will slow down the growth rate in the design of the IC size [1–5]. Within a microprocessor or a digital signal processor (DSP), information trail plays an important function because presentation metrics, viz. the die-area, speed of action, power dissipation and so on, directly affect the data path efficiency. As identified, the center of the data path involves composite computations like addition, subtraction, multiplication, division and so on. Thus, realizing well-organized hardware components intended for these calculations, which openly disturb the presentation of the statistics path, remains of major significance. The utmost performed process within the statistics path remains accumulation, which requires a binary adder just before it adds binary known numbers. Adders similarly show a dynamic part during additional composite calculations similar to exponentiation, division. Therefore, execution of a dualistic adder remains critical, aimed at an effective information path. Comparatively, important performance have been carried out during intending and then under studying well-organized adder paths intended for dual calculation. The present-day market of portable electronics, with battery power, needs a circuit plan with ultra-low-power dissipation. As the size along with the complexity of the chip continues to increase, low power is one of the major parameters to consider.

According to a study, which categorically analyzed the different power attributes in CMOS (Complementary Metal Oxide Semiconductor) circuits [6–9], there are active power, seepage power as well as short circuit power. The active power is the power used up for charging as well as discharging of the load capacitor. Cognitive radio and/or SDR (Software-Defined Radio) innately require multiband and multi-normal wireless circuits. The path is implemented based on the Si CMOS technology. In this chapter, the recent progress made in Si RF CMOS is described along with the reconfigurable RF CMOS circuit. In the future, more than a few kinds of the Si CMOS technology will be able to be worn for the RF CMOS circuit accomplishment. The sensible RF CMOS circuit performance just before cognitive and/or SDR is discussed. Recent progress of the Si CMOS LSI performance can enable the accomplishment of a GHz range RF circuit, so that the commercial wireless equipment are implemented using Si CMOS. A reconfigurable RF CMOS circuit is one hardware solution for SDR and/or cognitive radio [10].

In CMOS devices, very low leakage power dissipation occurs while entire inputs remain detained at various suitable logic stages and then the circuit is not within the indicated status. The total power dissipation occurs due to active power loss, small circuit power loss and leakage power loss, which can be represented by Eqs. (5.2–5.4), respectively.

$$\text{Total power dissipation} = P_{dynamic} + P_{short\text{-}circuit} + P_{Leakage} \quad (5.1)$$

whereas

$$P_{dynamic} = 0.5\ C_L V_{DD}{}^2 N_{0\text{--}1}\left(1/T\right) \quad (5.2)$$

$$P_{short\text{-}circuit} = I_{shot} \cdot V_{DD} \quad (5.3)$$

$$P_{leakage} = V_{DD} I_{leakage} \quad (5.4)$$

V_{DD} = power supply voltage.
N = switching activity.
C_L = load capacitance.
T = time period.

I_{shot} = short circuit current.

$I_{Leakage}$ = Leakage current.

5.2 Active Power Dissipation or Switching Power Dissipation

The active power dissipation is illustrated throughout an example of a CMOS inverter with a load capacitance C_L as shown in Figure 5.1a.

The CMOS inverter, which is in a steady state initially, will cause opposition toward the input voltage applied at the gate terminal. After the input signal makes the modification from HIGH to LOW, the PMOS transistor is ON and NMOS transistor is OFF, which causes the charging of the load capacitance C_L as shown in Figure 5.1b. On the other hand, the input signal changes starting LOW toward HIGH, the PMOS transistor is turned OFF and NMOS is turned ON, which creates a path for the discharge of load capacitor C_L described in Figure 5.1c [7–9].

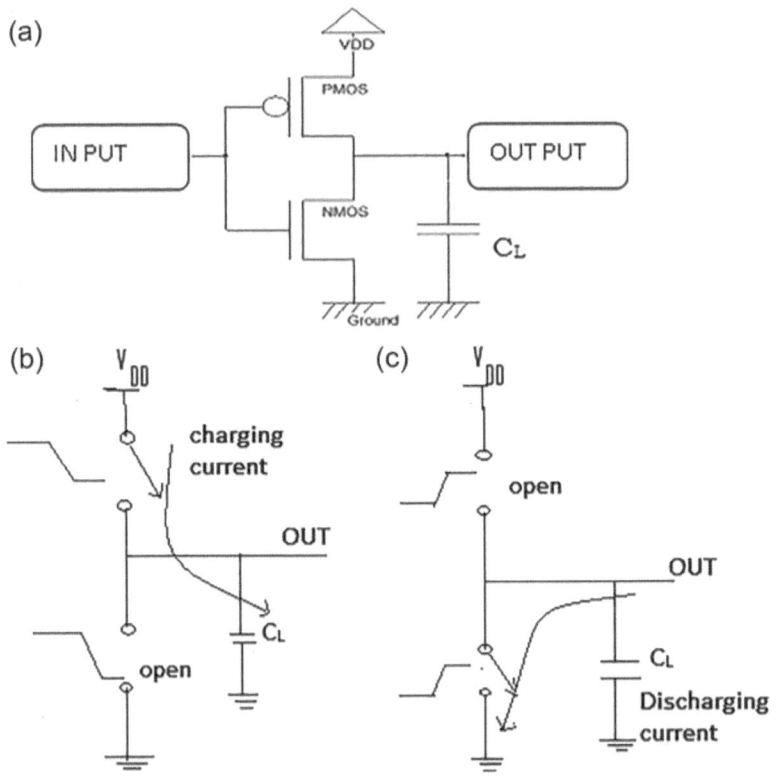

Figure 5.1 (a–c) The CMOS inverter.

The amount of growing transitions remains partial of the whole amount of transitions. Then Eq. (5.5) can be rewritten as follows:

$$P_{dynamic} = 0.5 \ C_L V_{DD}^2 N_{(0-1)}.f \tag{5.5}$$

In the above equation, $P_{dynamic}$ remains proportional to

- The load capacitance C_L.
- The square of V_{DD}.
- The swapping activity N.
- The timer frequency f.

As a result, the dynamic power drop is often completed in numerous ways:

- Decrease of load capacitance C_L.
- Decrease of the power source voltage.
- Decrease of clock frequency.

5.3 Short Circuit Power Dissipation

Short circuit power indulgence for CMOS circuits is found to occur in the transition period while NMOS and PMOS transistors together remain in ON condition temporarily. The short circuit current is particularly leading when the amount of the load capacitance produced remains small, or else as soon as the response signal's rise and fall times are large. Short circuit current remains important, as the increase/decrease time on the input of the gate is much larger than the output rise/fall time as shown in Figure 5.2.

$$P_{short-circuit} = I_{sc}.V_{DD} \tag{5.6}$$

5.4 Leakage Power Dissipation

In CMOS circuit power leakage happens in line for several conditions, then it may take place when the device output is not switching, the power due to the presence of reverse biased. A series of leakage current components and parasitic resistance affect the nature and magnitude of power loss. The power loss due to leakage is given in Eqs. (5.7–5.9), respectively.

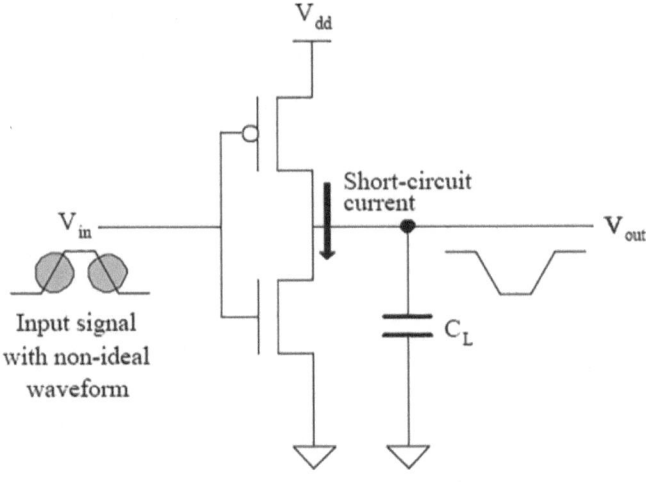

Figure 5.2   The operation of a CMOS inverter.

$$P_{leakage} = V_{DD}I_{leakage} \tag{5.7}$$

$$I_{Leakage} = V_{DD}/R \tag{5.8}$$

where $I_{Leakage}$ is the sum of various currents.

$$I_{leakage} = I_{rev} + I_{sub} + I_{DIBL} + I_{GIDL} + I_{PT} + I_{ox} + I_{HOT} \tag{5.9}$$

5.4.1 Junction Reverse Bias Leakage Current (I_{rev})

Junction reverse bias leakage current 'I_{rev}' can be identified as the reverse leakage current flowing from the source to drain of a MOS transistor. The motion of electron flow is caused by two conditions: (1) minority carrier flow as well as point close edge of the depletion region and (2) formation of electron–hole pairs at the depletion area of reverse biased connection, contributing near the general connection leakage current. The reverse current across the junction rests on the junction region followed by reverse saturation current density.

5.4.2 Subthreshold Conduction Leakage Current (I_{sub})

Subthreshold conduction leakage current I_{sub} flows when the supply voltage remains below the threshold voltage (V_{th}) across the source and drain. This stays more overnamed weak reversal current. These current movements in the OFF state in line with minority carriers populate the network through a diffusion procedure, after the gate-source voltage (V_{GS}) is situated fewer than V_{th}.

5.4.3 Drain-Induced Barrier Lowering Leakage Currents (I_{DIBL})

The initial power conditions in short channel devices significantly affect the subthreshold region current due to the lowering of the barrier in the depletion region. The drain biasing induces space-charge conditions in the short channel model, which changes drain biasing in the CMOS circuits. The observed leakage current, which changes drain biasing, is not seen in the case of long channel and change of threshold voltage and hence, it is known as the drain-induced barrier lowering leakage current.

5.4.4 Gate-Induced Drain Leakage Current (I_{GIDL})

Channel length modulation is the least affected by the threshold voltage and in the case of a long channel, the threshold voltage depends on short channel models. Popular short channel devices and parameters associated with the channel are determined by the source and drain space-charge region; hence, the subthreshold region current always varies with drain biasing.

5.4.5 Punch through Leakage Current (IPT)

When the source and drain region depletion parts come in close contact with the channel and the gate loses control over the channel current, it is known as the punch through condition, which varies the source to drain voltage.

5.4.6 Gate Oxide Tunneling Current (I_{ox})

Once there exists a presence of a strong electrical field transversing a thick oxide coating, there is an excavation of electrons due to outflow. Tunneling of charge current due to leakage is directly proportional to oxide thickness and voltage drop at the gate oxide due to the direct tunneling of electrons.

5.4.7 Hot Carrier Leakage Current (I_{HOT})

The transistor working in the saturated region will have a leakage current. The short channel effect causes the charged particles in the substrate region to gain sufficient energy from the field close to the silicon and oxide interface, which processes the oxide layer toward the gate. This effect is called hot carrier injection.

5.5 Proposed Theoretical Analysis of Parameters in CMOS Circuits

There are various types of CMOS circuits used in wireless communication principles include develop addicted to commercially existing and at present in use, e.g., WLAN, Bluetooth, GPS, DTV and RFID. The multi-mode movable phone happened to be admired in the concept of "cognitive radio," and wireless terminals communicate by means of the frequency band and/or the standard to not using db. For the CMOS 90 nm technology, varying theoretical parameters for CMOS circuits are shown in Tables 5.1–5.4.

1. NMOS transistor
2. PMOS transistor

Table 5.1   NMOS Transistor Vt

MOS CURRENT	NORMAL VT	HIGH VT	LOW VT
I_{on}	0.68 mA	0.63 mA	0.74 Ma
I_{off}	165 nA	30 nA	300 nA

Table 5.2   PMOS Transistor Vt

MOS CURRENT	NORMAL VT	HIGH VT	LOW VT
I_{on}	0.37 mA	0.35 mA	0.39 mA
I_{off}	78 nA	21 nA	135 nA

Table 5.3   Resistance Value of MOSFET

NMOS MOS CURRENT	NORMAL VT	HIGH VT	LOW VT
I_{on}	1.7 KΩ	1.9 KΩ	1.6 KΩ
I_{off}	22 MΩ	40 MΩ	4 MΩ
PMOS MOS Current	Normal vt	High vt	Low vt
I_{on}	3.2 KΩ	3.4 KΩ	3.0 KΩ
I_{off}	72 MΩ	57 MΩ	88 MΩ

Table 5.4   W/L Ratio of NMOS and PMOS

CHANNEL DIMENSIONS	NMOS (MM)	PMOS (MM)
W	1.2	3
L	0.1	0.1

Theoretical calculation of dynamic power

$$P_{total} = P_{Dynamic} + P_{Leakage} \tag{5.10}$$

$$P_{Dynamic} = 0.5V_{DD}^2.f.C.N \tag{5.11}$$

$$P_{leakage} = V_{DD}I_{leakage} \tag{5.12}$$

where
V_{DD} = power supply
f = clock frequency
C_L = load capacitance
N = switching activity

5.5.1 Calculation of CMOS NOT Gate Dynamic Power

The dynamic power for the CMOS gate is represented below [7–9].

$$P_{Dynamic} = 0.5V_{DD}^2.f.C.N$$

$$N = ?, C_L = ?$$

Considering NOT gate for the calculation of the dynamic power, the truth table is represented in Table 5.5.

N = switching activity
P = number of inputs (1)
Q = number of zeros entered in the output column (1)

$$N_{0-1} = Q.\left(2^P - Q\right)/\left(2^{2P}\right) \tag{5.13}$$

$$N_{0-1} = 0.25$$

Table 5.5   Truth Table of NOT Gate

INPUT	OUTPUT
0	1
1	0

5.5.2 Gate Capacitance per Unit Area (C_{ox})

$$Cox = \varepsilon_o \, \varepsilon_r / Tox, \qquad\qquad (5.14)$$

$\varepsilon_o = 8.854 * 10 - 12;$ (F/m) permittivity of medium
$T_{ox} = 2.\,05 * 10^{-9};$ effective oxide thickness (gate)
$\varepsilon_r = 3.9;$ relative permittivity of SiO_2

$$C_{ox} = \left(3.9 * 8.854 * 10^{-12}\right) / \left(2.05 * 10^{-9}\right) = 15 * 10^{-3} \text{ Farads}$$

5.5.3 Gate Capacitance (Cg)

$$Cg = \left(W.L.C_{OX}\right) \; {-}{-}{-} \, NMOS \qquad\qquad (5.15)$$

$$Cg = \left(1.2 * 10^{-6} * 0.1 * 10^{-6} * 15 * 10^{-3}\right),$$

$$Cg = \left(3 * 10^{-6} * 0.1 * 10^{-6} * 15 * 10^{-3}\right), \; {-}{-}PMOS$$

$$Cg = 1.8 * 10^{-15} F \; {-}{-}{-} \, NMOS \quad Cg = 5.4 * 10^{-15} F \; {-}{-}{-}PMOS$$

$$(3/2)\, Cg = 2.7 * 10^{-15} F {-}{-}{-}{-}NMOS$$

$$(3/2)\, Cg = 8.1 * 10^{-15} F - PMOS$$

$N = 0.25,$ VDD $= 1.2v,$
Total load capacitance $= 2.2$ Ff,
Time period $= 5$ns, $f = 1/T$

$$P_D = 0.5 * NfC_L V_{DD}{}^2$$

$$P_D = 0.5 * 0.25 * 2.2 * 10^{-15} * 0.200 * 10^9 (1.2)^2$$

$$P_D = 0.792 * 10^{-6} \text{Watts}$$

5.5.4 Leakage Power ($V_{DD}.I_{leak}$)

$$I_{leak} = V_{DD} / R$$

$$I_{leak} = 1.2 / 47 * 10^6$$

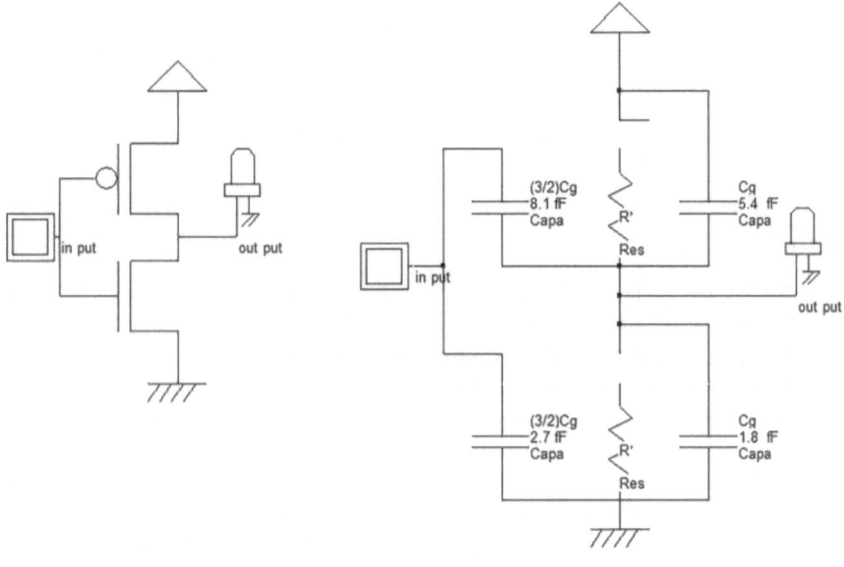

Figure 5.3 Resistance and capacitance of CMOS NOT gate.

Leakage power = $V_{DD}.I_{leak}$

$$P_{lea} = 1.2 * 0.025 * 10^{-6}$$

$$P_{lea} = 30 * 10^{-9} \, \text{Watts}$$

Example: 1-bit-CMOS full adder input and output truth table
  N = switching activity of full adder
  P = number of inputs (3)
  Q = number of zeros entered in the output column (4)

$$N_{0-1} = Q.(2^P - Q) / (2^{2P})$$

$$N_{0-1} = 4.(2^3 - 4) / 2^{2*3}$$

$$N_{0-1} = 16/64$$

$$N_{0-1} = 0.25$$

5.6 Conclusion

Herein, it is observed that the dynamic power may be reduced with the reduction of the load capacitance and the power supply voltage.

The reduction of the power increases the lifetime of the chip and performance calculated theoretical values. The RF CMOS technology for cognitive radio and/or SDR is discussed. It is pointed out that there are various approaches for reconfigurable RF CMOS circuits.

References

1. N. Weste and D. Harris, *CMOS VLSI Design: A Circuits and Systems Perspective*, Addison-Wesley, 4th ed. 2009.
2. D.K. Roy and S.C. Prasad, *Low-Power, CMOS VLSI Circuit Design*, John Wiley & Sons Inc., 2000.
3. S. Kang and Y. Leblebici, *CMOS Digital Integrated Circuits*, Tata McGraw- Hill, New York, 2003.
4. A. Bellauor and MohamahadIlmasrg, *Low Power Digital VLSI Design Circuits and Systems*, Kluwer Academic Publishers, 2nd ed.
5. J.P. Uyemura, *Introduction to VLSI circuit and systems*, John Wiley, 2002.
6. A.P. Chandrakasan, S. Sheng and Brodersen "Low power CMOS Digital Design." *IEEE Journal of Solid State Circuits*, vol 27, pp. 473–484, April 1992.
7. J. Brews, *High Speed Semi Conductor*, Wilky, New York 1990.
8. J. Rabaey, *Digital Integrated Circuits: A Design Perspective*, Prentice Hall, NJ, 1996.
9. E. Kamarn and S. Pucknell, *Essential of VLSI Circuits and Systems*, PHI, 2005.
10. K. Masu and K. Okada, "Reconfigurable RF CMOS circuit for cognitive radio" *IEICE Transactions on Communications* E91B, January 2008.

6

A Novel Design of 16 Bit MAC Unit Based on Vedic Mathematics Using FPGA Hardware for Cognitive Radio Application

DAYADI LAKSHMAIAH

Sri Indu Institute of Engineering and Technology

SHAIK FAIROOZ

Malla Reddy Engineering College (Autonomous)

FARHA ANJUM

Siddhartha Institute of Engineering and Technology

MOHAMMAD ILLIYAS

Shadan College of Engineering and Technology

I. SATYA NARAYANA

Sri Indu Institute of Engineering and Technology

Contents

DOI: 10.1201/9781003102625-6

6.1 Introduction

Vedic mathematics remained to reclaim in the mid-twentieth century since the old first nation models (Marai). Old Indian arrangement of mathematics stood forgotten since the Vedic sutras. The ordinary numerical calculations can be rearranged and also streamlined by the utilization of Vedic mathematics. The Vedic calculations can be applied to math, geometry, a basic in addition to round calculation and analytics. In Ref. [1], the authors have proposed another multiplier dependent on a Vedic calculation for low force and fast applications. Their multiplier architecture depends on creating every incomplete item and their wholes in a single step. They guarantee that their proposed Vedic multiplier is quicker than the exhibit multiplier and Booth multiplier. The authors in Ref. [2] have tried and analyzed different multiplier executions, for example, array multiplier, multiplier large scale, Vedic multiplier with full parceling, Vedic multiplier utilizing four-cycle full scale, completely recursive Vedic multiplier and Vedic multiplier utilizing eight-digit large scale for ideal speed. They have guaranteed that the Vedic technique is not on a very basic level and is not the same as the customary strategy for augmentation. The execution of Rivest, Shamir and Adleman (RSA) encryption/unscrambling calculation utilizing Vedic mathematics remains future to improve executions in Ref. [3]. They take the utilized Vedic multiplier and division construction in the RSA hardware for improved proficiency. Their outcomes show that RSA hardware actualized utilizing Vedic division and augmentation is productive as far as region/speed contrasted with its usage utilizing traditional duplication and division architectures. Dhillon and Mitra [4] proposed a multiplier utilizing the "Urdhva Tiryagbhyam" (UT) calculation, which is advanced by the "Nikhilam" calculation. They have recommended a decreased piece augmentation calculation utilizing the UT and "Nikhilam" sutras. We have built another Vedic multiplier structure utilizing the "Nikhilam" sutra. The carry-save viper executed in the proposed architecture diminishes propagation delay altogether. It is accepted that our architecture may set a new way for future exploration. The basic reconfigurable block for a cognitive radio implementation is the Vedic multiplier.

6.2 Literature Survey

6.2.1 "Hybrid Multiplier-based Optimized MAC Unit" Kavindra Dwivedi, R.K. Sharma and Ajay Chunduri

In any place where there is a requirement for superior processing applications, there is a clear interest in an effective rapid multiplier. Augmentation takes maximum huge time when contrasted with other number juggling tasks. Multipliers are the greatest fundamental squares in superior registering construction similar to numerical signal processors (DSP). A Macintosh element which comprises a multiplier and a collector accepts a significant part to choose the presentation of some DSP blocks [5–6]. The improved presentation of the Multiply Accumulator Circuit (MAC) element satisfies the boundary of rapid calculation and continuous handling abilities of DSP. Throughout, the long-term quantity of thoughts has been planned to recover the presentation and relieve the extreme fractional item time age by the customary augmentation method. Generally in this chapter, we take arranged zeroes offering the Mackintosh style utilizing an incorporated cross double multiplier, coordinated CLA viper organization. The incorporated multiplier is a blend of Karatsuba calculation and UT sutra after Vedic arithmetic. The CLA viper system comprises CLA and a restrictive aggregate snake, which assists with decreasing time by performing equal expansion. The referenced plan is executed in Verilog HDL utilizing a Libero SOC PolarFire v2.1 device, focusing on its Polar Fire FPGA personal besides the MPF300T_ES-1FCG484E gadget.

6.2.2 *"Parallel Multiplier Stake Element Built On Vedic Calculation" Jithin S. and Prabhu E.*

Now in this chapter, a productive equal multiplier besides a collector (Mackintosh) unit dependent on Vedic mathematics is introduced. Vedic math uses the UT sutra aimed at the multiplier design. The projected Mackintosh design elevates the quickness of activity by lessening the entryway region and by force scattering. We can likewise accomplish a better stay by the assistance of the Vedic encoder tracked via the evacuation of the gatherer phase by parallelizing the moderate outcomes, taking care of the info. Such pipelining of the halfway outcomes, before the last viper, takes the impact of joining the aggregator phase through the fractional item phase of the multiplier. Furthermore, the general calculation rapidity of the Mackintosh part remains raised by the effective utilization of higher request blowers in the consolidated fractional item pressure and collector (PPCA) architecture [7]. The region, effectiveness and force intelligence display that, the basic way stay of the planned project is altogether diminished and it beats the current projects. There is an increment of 20%–30% and 7%–18% separately aimed at the 4-cycle then eight-digit Vedic Mackintosh units, regarding the situation all out route power, basic way interruption and chamber territory. The construction was orchestrated utilizing the usual 90 nm CMOS collection and actualized on Altera's Hurricane II arrangement FPGA.

6.2.3 "Proposal in Addition Optimization of 16×16 Bit Multiplier by Vedic Arithmetic" Sheetal N. Gadakh and Amitkumar Khade

Augmentation is a fundamental capacity in number juggling activities. Augmentation-based activities, for example, duplicate and Accrue component (MAC), convolution and fast Fourier transform (FFT) are generally utilized for trendy indication handling applications. As duplication rules the implementation season of DSP frameworks, it is essential for growing fast multipliers. Old Vedic arithmetic encourages the answer to a certain degree [8]. Now in this chapter, the idea of UT is utilized, i.e., perpendicularly and transverse duplication to execute 16×16 Bit Vedic multiplier and the advancement remain accomplished using carry bar adders. Contrasting and past constructions, planned manner accomplishes 33.26% decrease in combinational way interruption. The Vedic multiplier suggests dstands actualized in VHDL while incorporated and re-enacted utilizing Xilinx ISE Project Suite 14.5.

6.2.4 "Proposal of Mackintosh Element in Artificial Neural System Construction with Verilog HDL" L. Ranganath, P.G. Scholor, D. Jay Kumar and P. Siva Nagendra Reddy

An artificial neural union (ANU) is the same as a data preparing construction comprised of handling components. The preparing unit chooses though the organization stands effective before or not, therefore to project a proficient preparing element and additionally give better execution. The preparing unit comprises a MAC element (development and accumulation) and an activation element. Now for a current framework, the preparing MAC element is intended in Booth multiplier and carry look forward snake. The current preparing component gives delay and burns through an extra region and force. To overcome the disadvantages, design another handling element, Vedic multiplier through square root carry select viper (SQRT-CSLA). The projected plan beats the downsides of the current framework, also it is likewise giving an improved presentation of the whole organization. The beginning effort element remained intended by sigmoid neurons measure. The whole handling element remained actualized and confirmed through utilizing Verilog HDL language.

6.3 Existing Method Vedic Arithmetic

Vedic arithmetic is an exceptionally ancient method that is straightforwardly used in different parts of arithmetic, for example, polynomial math, number-crunching and so on [9–10]. It diminishes the multifaceted nature by eliminating the pointless advances while ascertaining any outcome. Here are 16 sutras in Vedic maths. The accompanying track-down displays the rundown of every single Vedic sutra.

1. (Anurupye) Shunya anyat
2. Antyayoreva
3. Anurupyena
4. Ekanyunena Purvena
5. Gunakasamuccayah
6. Gunitasamuccayah
7. Nikhilam
8. Paraavartya
9. Puranapuranabyham
10. Sankalana
11. Shesanyankena
12. Shunyam-Saamyasamuccaye
13. Sopantyadvayamantyam
14. Urdhva Tiryakbhyam
15. Vyashtisamastih
16. Yavadunam

Besides the 16 sutras, Adyamadyenantyamantyena and Nikhilam Navatashcaramam Dashatah (NND) are utilized aimed at registering growth of any binary quantities. For the most part, the NND sutra stands liked on behalf of bigger piece quantities and the UT sutra remains liked aimed at more modest piece numbers. Consequently, the UT sutra is utilized now in this effort. UT signifies "vertically and crosswise" [4]. This one stands utilized aimed at an increase of binary quantities through one base. Allowing us to consider the system for duplicating two 3-cycle statistics say U [2:0] and V [2:0] for instance and C [3:0] indicates the carry, and Y [2:0] means the halfway item yield. At that point, the accompanying advances remain toward surveyed (Figures 6.1 and 6.2):

step1: C0Y0 = U0V0

step2: C1Y1 = {(U0*V1) + (U1*V0)} + C0

step3: C2Y2 = {(U0*V2) + (U1*V1) + (U2*V0)} + C1

step4: C3Y3 = {(U1*V2) + (U2*V1)} + C2

step5: C4Y4 = {(U2*V2)} + C3

Hereafter, absolute product = C4Y4Y3Y2Y1Y0

So 21 × 32 = 276.

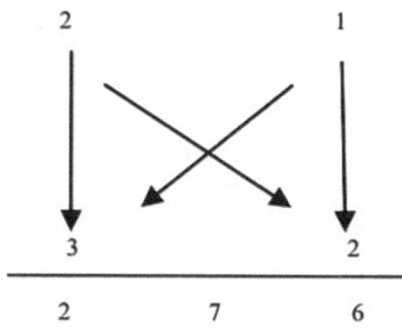

Figure 6.1   Multiplication for 21×32.

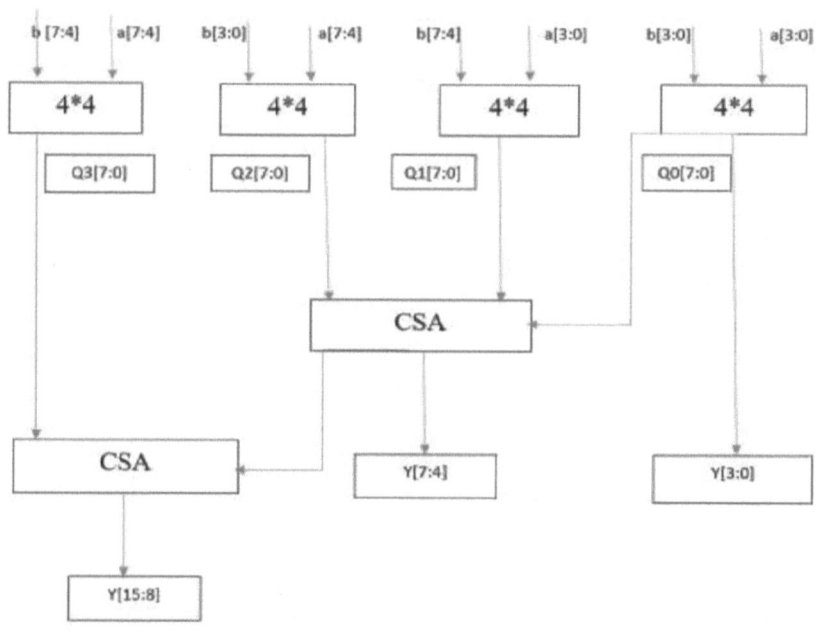

Figure 6.2   Further multiplication.

6.3.1 Straightforward Multiply Accumulator Circuit (MAC)

Now the proposal of a 16-digit MAC element, eight cycle multiplier is utilized. This is intended by 4-bit multipliers. Also, a 4-bit multiplier is planned to utilize 2-bit multipliers. The figure displays MAC. The usage of the 2-digit multiplier with half adders utilizing UT sutra is given in Figure 6.3.

6.3.2 Carry-Save Adder

It aims at the most parts utilized for processing expansion of at least 3 n cycle numbers. Herein, the three sources of info are changed over to two yields where one yield indicates fractional total and the additional one addresses carry. The last total is given by moving the transfer toward leftward by one cycle place and then afterward affixing the MSB of incomplete entirety through zeroes (Figures 6.4–6.8; Table 6.1).

Top Level Module:

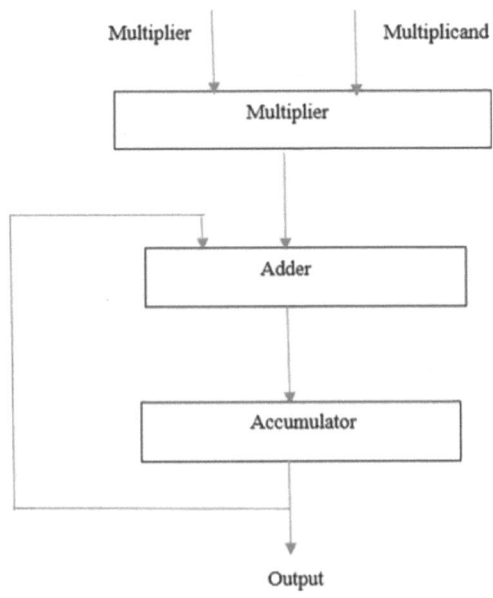

Figure 6.3 Basic MAC.

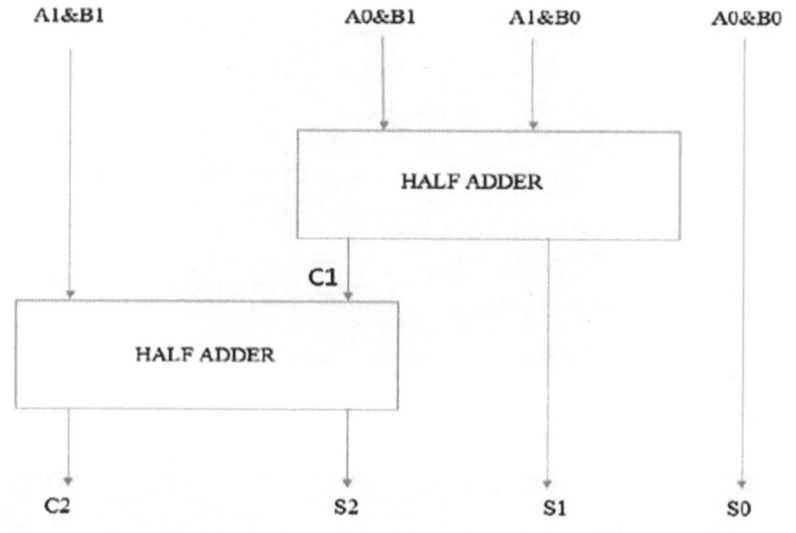

Figure 6.4   Implementation of a 2-bit multiplier through partial adders.

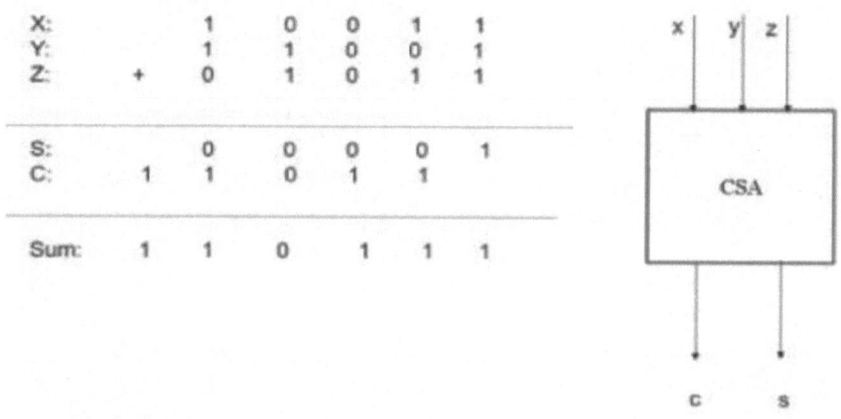

Figure 6.5   Carry-save adder.

6.4 Proposed Method and Performance and Simulation Results

6.4.1 Adders

In hardware, a snake stands for a numerical circuit that executes expansion. Indeed, the duplication activity relies upon the arrangement of expansion activity.

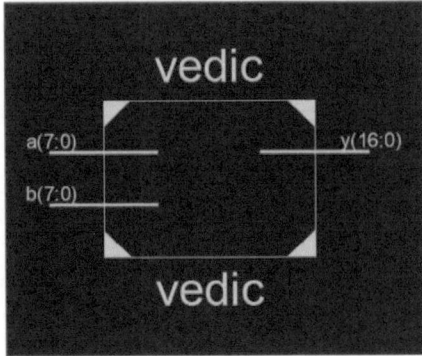

Figure 6.6 RTL schematic (existing).

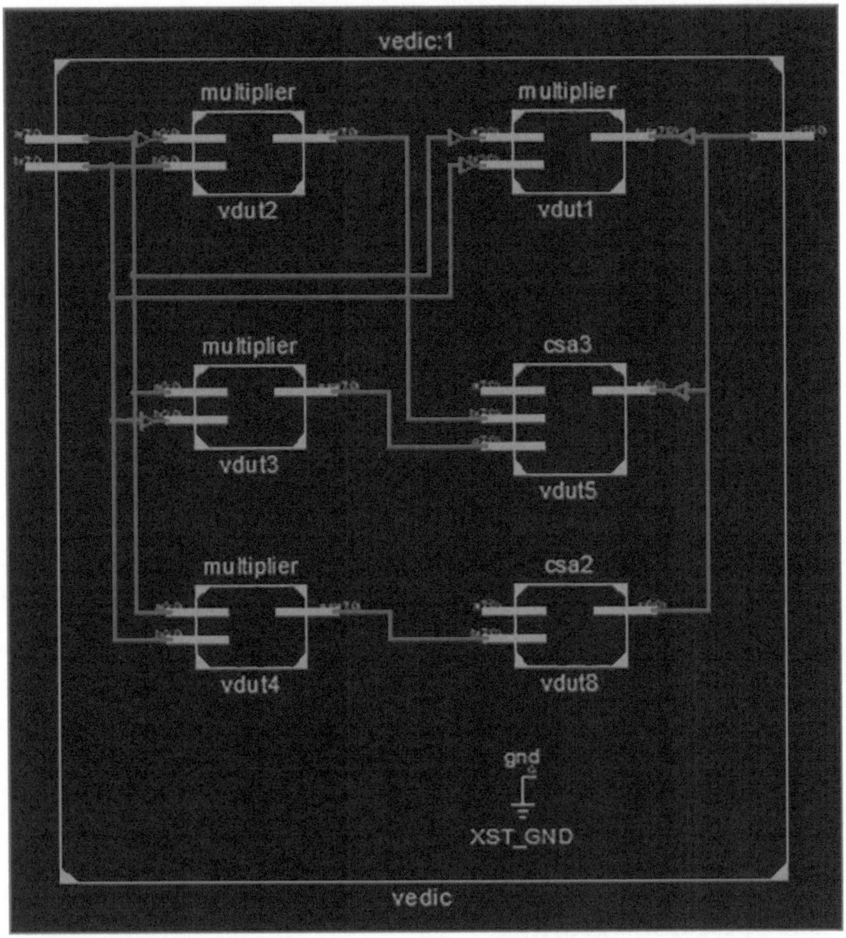

Figure 6.7 Output (existing).

vedic Project Status (02/10/2020 - 17:18:36)			
Project File:	UT.xise	Parser Errors:	No Errors
Module Name:	vedic	Implementation State:	Synthesized
Target Device:	xc6slx9-3tqg144	• Errors:	No Errors
Product Version:	ISE 14.7	• Warnings:	No Warnings
Design Goal:	Balanced	• Routing Results:	
Design Strategy:	Xilinx Default (unlocked)	• Timing Constraints:	
Environment:	System Settings	• Final Timing Score:	

Device Utilization Summary (estimated values)				[-]
Logic Utilization	Used	Available	Utilization	
Number of Slice LUTs	121	5720	2%	
Number of fully used LUT-FF pairs	0	121	0%	
Number of bonded IOBs	33	102	32%	

Figure 6.8   Device utilization summary (existing).

Table 6.1   Summary Report Existing Work

VEDIC PROJECT STATUS (02/10/2020-13:32:46)

Project file:	UT_KSA.xise	Parser errors:	No errors
Module name:	Vedic	Implementation state:	Synthesized
Target device	Xc6slx9-3tqg144	• Errors:	
Product version:	ISE 14.7	• Warnings:	
Design goal:	Balanced	• Routing results:	
Design strategy:	Xilinx default (unlocked)	• Timing constraints:	
Environment:	System settings	• Final timing score:	

DEVICE UTILIZATION SUMMARY (ESTIMATED VALUES)

Logic Utilization	Used	Available	Utilization
Number of slice LUTs	121	5720	2%
Number of fully used LUT-FF pairs	0	121	0%
Number of bonded IOBs	33	102	32%

Adders can be actualized in various protocols utilizing various advancements at various degrees of architecture. A proposal of great velocity and dependable calculators is the excellent target and prerequisite aimed at installing requests besides separating activity additionally utilizing adders.

Table 6.2 Timing Analysis Report Existing Work

```
Data Path: a<2> to y<15>
                            Gate    Net
   Cell:in->out   fanout   Delay   Delay   Logical Name (Net Name)
 --------------   ------   -----   -----   -------------
   IBUF:I->O         21    1.222   1.218   a_2_IBUF (a_2_IBUF)
   LUT2:I0->O         2    0.203   0.981   vdut1/p91 (vdut1/p9)
   LUT6:I0->O         3    0.203   0.651   vdut1/g26/hcarry1 (vdut1/c3)
   LUT4:I3->O         3    0.205   0.879   vdut1/g27/g191_xo<0>1 (vdut1/w2)
   LUT6:I3->O         2    0.205   0.961   vdut1/g29/hcarry1 (vdut1/c6)
   LUT6:I1->O         3    0.203   0.879   vdut1/g30/g191_xo<0>1 (vdut1/w4)
   LUT6:I3->O         3    0.205   0.651   vdut1/g32/hcarry1 (vdut1/c9)
   LUT3:I2->O         2    0.205   0.617   vdut1/g33/g191_xo<0>1 (vdut1/w6)
   LUT5:I4->O         3    0.205   0.651   vdut1/g34/g191_xo<0>1 (yl<5>)
   LUT5:I4->O         3    0.205   1.015   vdut5/c3dut2/Madd_n0003_Madd_lut<0>1 (vdut5/c3dut
   LUT6:I0->O         3    0.203   0.995   vdut5/c3dut9/Madd_n0003_Madd_lut<0>1 (y_5_OBUF)
   LUT6:I1->O         2    0.203   0.981   vdut5/c3dut10/Madd_n0003_Madd_cy<0>11 (vdut5/c3du
   LUT6:I0->O         3    0.203   1.015   vdut5/c3dut11/Madd_n0003_Madd_cy<0>11 (vdut5/c3du
   LUT6:I0->O         3    0.203   1.015   vdut8/cdut12/Madd_n0004_lut<0>1 (y_8_OBUF)
   LUT6:I0->O         2    0.203   0.864   vdut8/cdut92/Madd_n0004_cy<0>11 (vdut8/cdut92/Mad
   LUT4:I0->O         3    0.203   0.898   vdut8/cdut102/Madd_n0003_Madd_cy<0>11 (vdut8/cdut
   LUT5:I1->O         4    0.203   1.028   vdut8/cdut122/Madd_n0003_Madd_lut<0>1 (vdut8/cdut
   LUT6:I1->O         1    0.203   0.579   vdut8/cdut152/Madd_n0003_Madd_lut<0>1 (y_15_OBUF)
   OBUF:I->O                2.571           y_15_OBUF (y<15>)
 --------------------------------------------
   Total                  23.137ns (7.256ns logic, 15.881ns route)
                                   (31.4% logic, 68.6% route)
```

6.4.2 Proposed Block Diagram

KSA is an equal preface structure transmitting aspect forward snake; it is broadly measured by the way of the quickest viper and is generally utilized in the business for superior number-crunching circuits. The total working of KSA can be handily grasped by breaking it down into three unmistakable parts. Here, the remaining three phases of the calculation in PPA are given below (Figure 6.9).

- Pre-processing
- Prefix
- Ending computation

6.4.3 Pre-Processing

$P_i = A_i$ XOR B_i $G_i = A_i$ AND B_i Prefix:

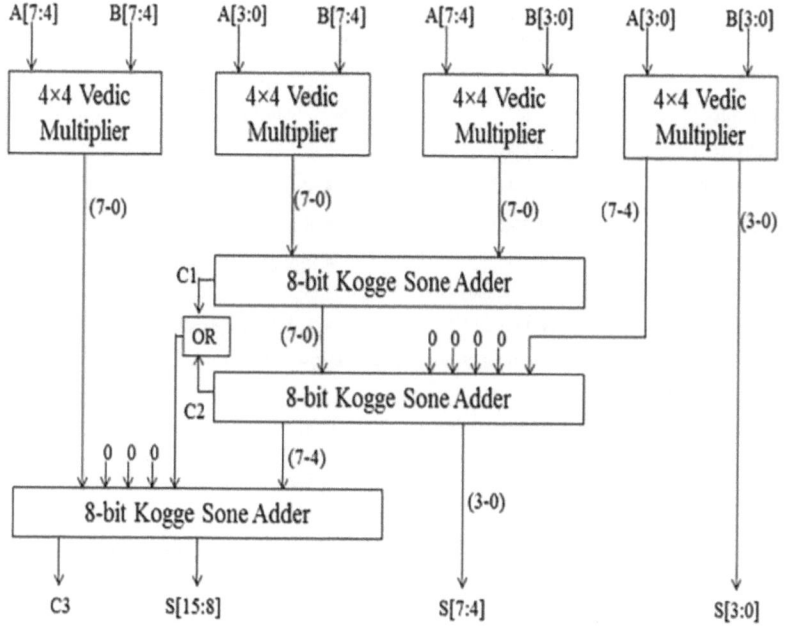

Figure 6.9 Proposed adders.

The dark chamber proceeds dual sets of produce besides spread indications (Gi, Pi) and (Gj, Pj) by way of info and figures a couple of creating and proliferating signs (G, P) as yield

G=Gi OR (Pi AND Pj) P=Pi AND Pj

6.4.4 Gray Cell

The dark cell proceeds two sets of creating and spreads signals (Gi, Pi) and (Gj, Pj) as information sources and registers a produce signal G, as yield.

G=Gi OR (Pi AND Pj)

6.4.5 Last Computation

This one includes control of whole pieces. Total pieces are processed through the rationale given beneath $Si = Pi$ XOR $Ci\text{-}1$ (Tables 6.3–6.5).

Table 6.3 Proposed Device Utilization

VEDIC PROJECT STATUS (02/10/2020-13:32:46)			
Project file:	UT_KSA.xise	Parser errors:	No errors
Module name:	Vedic	Implementation state:	Synthesized
Target device	Xc6slx9-3tqg144	• Errors:	
Product version:	ISE 14.7	• Warnings:	
Design goal:	Balanced	• Routing Results:	
Design strategy:	Xilinx default (unlocked)	• Timing Constraints:	
Environment:	System settings	• Final Timing Score:	

DEVICE UTILIZATION SUMMARY (ESTIMATED VALUES)			
Logic Utilization	Used	Available	Utilization
Number of slice LUTs	10	5720	0%
Number of fully used LUT-FF pairs	0	10	0%
Number of bonded IOBs	24	102	23%

Table 6.4 Proposed Device Utilization

```
Delay:              20.338ns (Levels of Logic = 17)
  Source:           a<2> (PAD)
  Destination:      y<16> (PAD)

Data Path: a<2> to y<16>
                            Gate    Net
    Cell:in->out    fanout  Delay   Delay  Logical Name (Net Name)
    ------------------------------------   ------------
    IBUF:I->O         18    1.222   1.154  a_2_IBUF (a_2_IBUF)
    LUT2:I0->O         2    0.203   0.981  vdut1/p91 (vdut1/p9)
    LUT6:I0->O         3    0.203   0.651  vdut1/g26/hcarry1 (vdut1/c3)
    LUT4:I3->O         3    0.205   0.879  vdut1/g27/g191_xo<0>1 (vdut1/w2)
    LUT6:I3->O         2    0.205   0.961  vdut1/g29/hcarry1 (vdut1/c6)
    LUT6:I1->O         3    0.203   0.879  vdut1/g30/g191_xo<0>1 (vdut1/w4)
    LUT6:I3->O         3    0.205   0.651  vdut1/g32/hcarry1 (vdut1/c9)
    LUT3:I2->O         2    0.205   0.617  vdut1/g33/g191_xo<0>1 (vdut1/w6)
    LUT5:I4->O         2    0.205   0.864  vdut1/g34/out1 (vdut1/c11)
    LUT6:I2->O         2    0.203   0.961  vdut1/g35/g191_xo<0>1 (y1<6>)
    LUT6:I1->O         3    0.203   0.995  vdut6/ci<2>1 (vdut6/ci<2>)
    LUT6:I1->O         3    0.203   1.015  vdut6/ci<4>1 (vdut6/ci<4>)
    LUT6:I0->O         3    0.203   0.651  vdut6/ci<6>1 (vdut6/ci<6>)
    LUT4:I3->O         2    0.205   0.961  c1 (c)
    LUT5:I0->O         4    0.203   0.684  vdut8/ci<4>1 (vdut8/ci<4>)
    LUT3:I2->O         1    0.205   0.579  vdut8/Mxor_si<6>_xo<0>1 (y_14_OBUF)
    OBUF:I->O               2.571          y_14_OBUF (y<14>)
    ------------------------------------
    Total             20.338ns (6.852ns logic, 13.486ns route)
                               (33.7% logic, 66.3% route)
```

Table 6.5 Proposed Summary Report

VEDIC PROJECT STATUS (01/28/2021-14:01:29)			
Project file:	UT_KSA.xise	Parser Errors:	No Errors
Module name:	Vedic	Implementation State:	Synthesized
Target device	Xc6slx9-3tqg144	• Errors:	
Product version:	ISE 14.7	• Warnings:	
Design goal:	Balanced	• Routing Results:	
Design strategy:	Xilinx Default (unlocked)	• Timing Constraints:	
Environment:	System Settings	• Final Timing Score:	
DEVICE UTILIZATION SUMMARY (ESTIMATED VALUES)			
Logic Utilization	Used	Available	Utilization
Number of slice LUTs	10	5720	0%
Number of fully used LUT-FF pairs	0	10	0%
Number of bonded IOBs	24	102	23%

DETAILED REPORTS					
Report Name	Status	Generated	Errors	Warnings	Infos
Synthesis report	Current	Mon 10,Feb 12:56:42 2020			
Translation report					
Map report					
Place and route report					
Power report					
Post-PAR static timing report					

6.4.6 Proposed Diagram

6.4.7 Hardware Requirement

6.4.7.1 General Integrated Circuits

The term coordinated circuit is utilized to depict a wide assortment of gadgets going from basic rationale entryways to complex cutting edge microprocessors. Coordinated circuits essentially comprise a circuit, ordinarily made up of various semiconductors and their interconnections, manufactured from a solitary semiconductor chip or bite the dust (Figures 6.10–6.13).

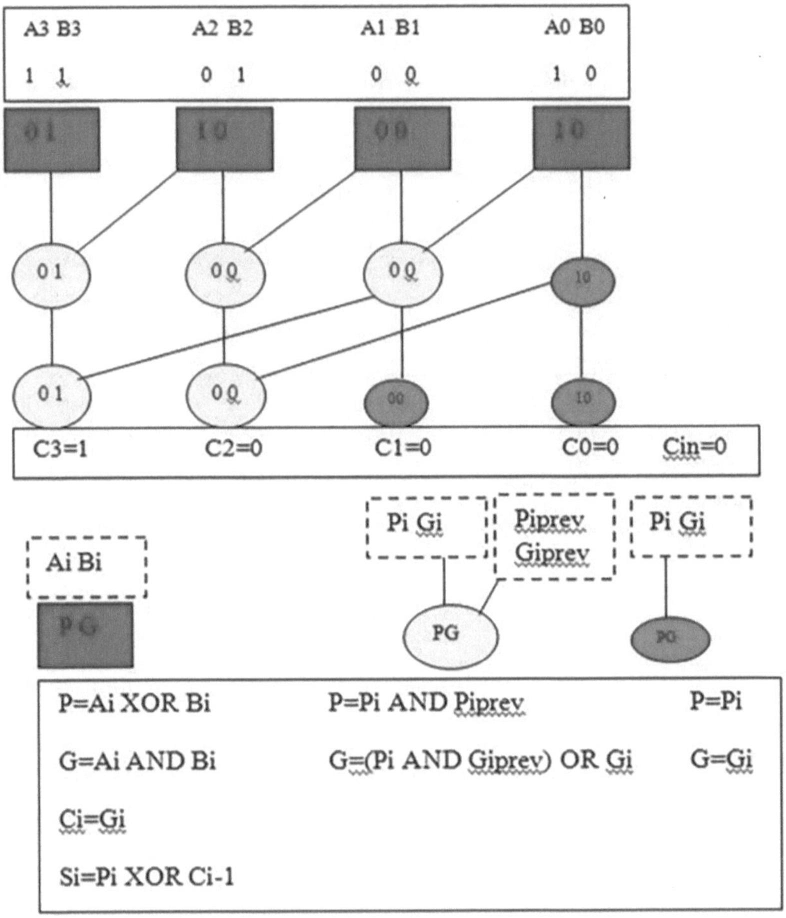

A3 B3	A2 B2	A1 B1	A0 B0
1 1	0 1	0 0	1 0

C3=1	C2=0	C1=0	C0=0	Cin=0

Ai Bi

Pi Gi | Piprev Giprev | Pi Gi

P=Ai XOR Bi	P=Pi AND Piprev	P=Pi
G=Ai AND Bi	G=(Pi AND Giprev) OR Gi	G=Gi
Ci=Gi		
Si=Pi XOR Ci-1		

Figure 6.10   Proposed top-level module.

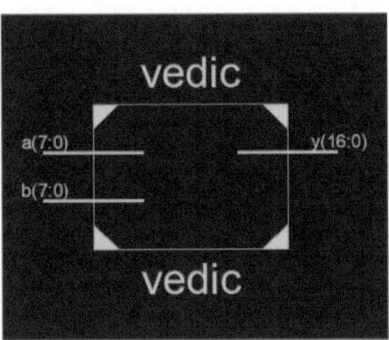

Figure 6.11   Proposed RTL schematic 1.

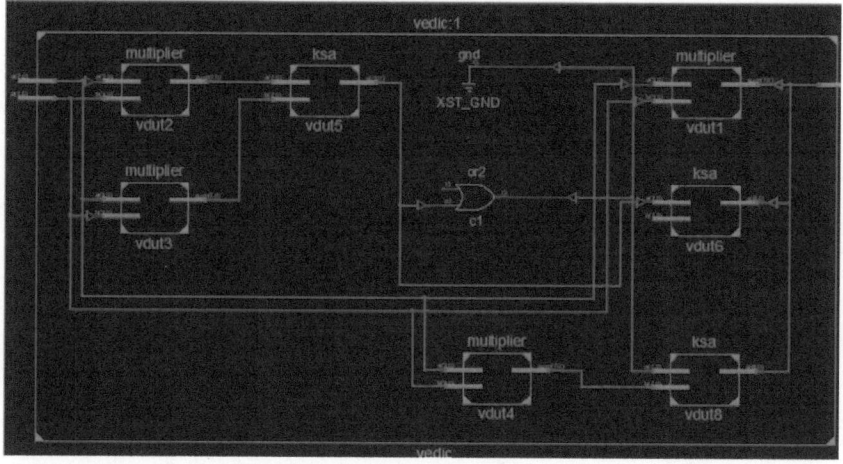

Figure 6.12   Proposed RTL schematic 2.

Name	Value	0 us	1 us	2 us	3 us	4 us
a[7:0]	z	z	25	45	39	95
b[7:0]	z	z	25	32	86	50
y[16:0]	x	X	625	1440	3354	4750
y1[7:0]	x	X	81	0	42	30
y2[7:0]	x	X	9	26	35	45
y3[7:0]	x	X	9	0	12	10
y4[7:0]	x	X	1	4	10	15
s1[8:0]	x	X	18	26	47	55
s2[8:0]	x	X	23	26	49	56
c	x	X				

Figure 6.13   Timing analysis (proposed system).

a. Analog coordinated circuits

Simple coordinated circuits incorporate a wide scope of utilizations, large numbers of which are exceptionally explicit. A few models are the basic operational speakers and clocks, and the more mind-boggling FM sound system decoders and single-chip FM radios.

There has been a pattern toward manufacturing the more usually utilized simple circuits into a single-chip structure.

An illustration of this is FM radio beneficiary, which is a genuinely unpredictable circuit when manufactured from discrete segments. An FM radio recipient would now be able to be developed from an FM radio chip, a sound speaker chip and a couple of discrete latent parts.

b. Computer coordinated circuits

PC coordinated circuits are gadgets, which structure the dynamic segments of a PC framework. They are regularly utilized related to digital coordinated circuits, which give a 'stick rationale' work. PC coordinated circuits can be practically partitioned into microprocessors, memory gadgets and fringe control gadgets.

6.5 Conclusion

A 16-digit MAC element using an eight-cycle Vedic multiplier with transmitting apart from snake remained intended. This depends on the UT sutra and implies the popular Verilog HDL. The usage remains approved happening Artix-7 FPGA. This one remains to be seen toward having about 9.5% force decrease alongside critical improvement in zone and delay. An examination with an ordinary multiplier and the current multiplier remained completed separately. The MAC element intended through the planned multiplier can be utilized in DSP presentations for refining the quickness and execution. Now upcoming, this effort can be stretched out through supplanting the multipliers using reversible rationale doors aimed at accomplishing additional force decrease besides good execution.

- By comparing both the existing Vedic multiplier and proposed Vedic multiplier, it is seen that the proposed Vedic multiplier streamlines the delay.
- So, it is inferred that a multiplier that requires quick usage can utilize this kind of multiplier in picture and signal preparing applications. A new optimized flexible sensing scheme is used to reduce the area and complexity by considering a simple filter structure using D flip flops, multipliers and adders. The area, power and delay constraints are observed using the XILINX ISE simulator. The optimized flexible sensing scheme offers

better performance with a reduced area when compared to the flexible spectrum sensing technique thus making it a more simple and efficient structure. In the future, the work can be extended by implementing this technique using cooperative spectrum sensing. This sensing scheme can also be implemented using other novel detection techniques like Eigen value-based detection to obtain better performance at a low signal-to-noise ratio (SNR).

References

1. K. Dwivedi "Hybrid Multiplier-Based Optimized MAC Unit", *2018 9th International Conference (ICCCNT)*, Bangalore, 2018, pp. 1–4

2. Jithin "Parallel Multiplier – Accumulator Unit based on Vedic Mathematics", in *ARPN Journal of Engineering and Applied Sciences*, May 2015.

3. S. N. Gadakh "Design and Optimization of 16×16 Bit Multiplier Using Vedic Mathematics", in *Proc. Int. Conf. on Automatic Control and Dynamic Optimization Techniques*, 2016.

4. K. Bathija "Low Power High Speed 16x16 bit Multiplier using Vedic Mathematics", in *International Journal of Computer Applications*.

5. M. V. Durga Pavan "An Efficient Booth Multiplier Using Probabilistic Approach", in *2018 International Conference on Communication and Signal Processing (ICCSP)*, Chennai, 2018.

6. L. Ranganath "Design of MAC Unit in Artificial Neural Network Architecture using Verilog HDL", in *International Conference on Signal Processing, Communication, Power and Embedded System (SCOPES)*, 2016.

7. Carry-Save Addition. Prof. Loh. CS3220 – Processor Design – Spring 2005.

8. R. Balakumaran "Design of High Speed Multiplier Using Modified Booth Algorithm with Hybrid Carry Look-Ahead Adder", in *2016 International Conference on Circuit, Power and Computing Technologies (ICCPCT)*, Nagercoil, 2016.

9. A. Eshack "Implementation of Pipelined Low Power Vedic Multiplier", in *2nd International Conference on Trends in Electronics and Informatics (ICOEI)*, Tirunelveli, 2018.

10. S. Rakesh "A Survey on the Design and Performance of Various MAC Unit Architectures", in *2017 IEEE International Conference on Circuits and Systems (ICCS)*, Thiruvananthapuram, 2017.

7

HYBRID OPTIMIZATION TECHNIQUE USING PARTICLE SWARM OPTIMIZATION AND FIREFLY ALGORITHM USING THE CHORD PROTOCOL

DAYADI LAKSHMAIAH, R. YADAGIRI RAO, A. SINDHUJA, AND SURESH BALLALA

Sri Indu Institute of Engineering and Technology

P. PRASANA MURALI KRISHANA

KITS College

Contents

7.1 Introduction

Chord remains an organized P2P (peer-to-peer) algorithm that employs reliable chopping toward constructing a distributed hash table in a network consisting of numerous nodes. With the help of a consistent hash algorithm, say SHA-1, chord assigns m-bit identifiers to both peers and data items. The peers are arranged in a virtual ring (chord ring) according to their identifiers. The identifier of a data item

DOI: 10.1201/9781003102625-7

is known as a resource ID or key. A key is assigned near a knob whose identifier remains better than or else equivalent near it. In chord, there is always a uniform distribution of keys among all participating peers. Every participant continues a direction-finding table also named as an extremity bench that consists of routing information of log N other peers in an N-node network. The chord runs a stabilization algorithm to keep the finger tables of each peer up-to-date. The basic lookup for a node N searching for a resource K in a chord ring is given as follows:

Step1: The node N searching for a resource K first checks in itself. If it finds the key, then the search becomes successful and it returns the key. Otherwise, go to the next step.

Step2: The node N checks its finger table to find a node N' that is closer to key K but less than or equal to K. It then sends the search message to node N'. Step2 is repeated until the search becomes successful.

The original chord lookup method implemented in a WMN not only increased the number of lookup messages in the network but also took up the valuable bandwidth resources and led to congestion on the wireless links [1]. This method also leads to an increased delay in searching and answering a query. Therefore, it is important to reduce the query response time and traffic in the network to facilitate the well-organized source distribution in P2P established wireless mesh networks. A routing metric helps to overcome the query response time and traffic problem where the direction-finding procedures choose tracks that steady circulation capacity altogether beside movement track, and minimizes the inter-flow interfering required happening entire neighboring site. The investigation of the chord protocol over wireless mesh networks in Chapter 4 demonstrates that the performance of the chord protocol degrades due to the dynamic nature of nodes. The objective of this effort is to decrease the query response period, increase the packet delivery ratio and reduce the network traffic that interns network load for enabling well-organized source allocation of now P2P-constructed wireless mesh networks. Since the problem is to improve multiple QoS parameters, it is considered an optimization problem. The MESH-DHT discussed in Chapter 5 considered the connection superiority, end-to-end interruption, demand reaction period and package distribution percentage in addition to the existing location-aware chord protocol. The atom swarm optimization algorithm works toward improving the QoS parameters. The simulation consequences are shown. It is demonstrated that the planned MESH-DHT approach improved the presentation of chord efficiently when likened through an existing chord to an existing location-aware chord.

7.2 Proposed Methodology

This section discusses firefly algorithm (FFA) and cross of particle swarm optimization in addition to firefly algorithm (PSO-FF) toward improving QoS limits in P2P-founded wireless mesh networks.

7.2.1 Firefly Algorithm

The QoS parameters, now a P2P-centered wireless mesh network, form a non-deterministic polynomial though problematic, hereafter considered as an optimization problem. Nature encouraged meta-heuristic algorithms such as FFA, particle group optimization, ant optimization and so on have drawn considerable attention in the recent past to solve the optimization problems. Any optimization problem consists of an objective function. The optimization here is to find the best limits that minimize this purpose; otherwise, it maximizes this purpose. Herein, the objective is to consume small end-to-end interruption, great package distribution part, low netload and little typical demand response time in a P2P-based wireless mesh network.

The FFA is a nature stimulated meta-heuristic algorithm established by Xin-She Yang and was developed based on the idealized flashing performance of fireflies [2]. It remains an original ecology intelligence meta-heuristic algorithm used for solving various optimization problems. The additional fireflies communicate to each other, search aimed at request the nentice additional fireflies (especially, the opposite sex fireflies) by using bioluminescence through different flashing patterns. By mimicking nature, numerous meta-heuristic procedures are planned. The phenomenon of bioluminescence and blinking behavior of fireflies is the foundation for the FFA.

The major critical issues in the FFA are the preparation of drawing and modification of bright strength. Aimed at ease, certain blinking features of fireflies remain perfect for the firefly-inspired algorithm. The three perfect guidelines are defined as follows [3]:

1. Fireflies are assumed to be of the same sex. Therefore, one firefly remains fascinated toward another despite their gender.
2. The attraction between two fireflies remains proportionate toward their illumination that drops with an increase in detachment among fireflies. Aimed at every two given fireflies, the lesser bright firefly will move toward a brighter firefly. If a firefly does not find a firefly brighter than it, its drive changes aimlessly in the search space.
3. The illumination of a firefly remains strong-minded or influenced by its objective purpose.

The major advantages of the FFA, when compared with other heuristic algorithms, are the automatic sub-division of population and capability toward transaction through multi-modality. The highest FFA is found besides attractiveness between two fireflies in contrary wise relative near distance. Due to the multi-modality ability, the entire population is sub-divided into subgroups automatically, and each subgroup swarms everywhere a method or else is narrow optimal. The top worldwide result can initiate all these modes. Next, this section will allow the fireflies in the direction of identifying all goals instantaneously even though the uncertainty population scope remains suitably larger than the number of ways [4].

In a continuous optimization problem with a cost function say f, the objective is to reduce this total function f (as shown in Eq. (7.1)).

$$f(x') = \min_{x \in s} f(x) \tag{7.1}$$

In firefly-based optimization algorithms, the problem starts with the initial number of fireflies. Each firefly represents a potential solution for the proposed problem. The illumination or light strength of every firefly stays considered through the independent or cost purpose. In a simple case of an optimization problem, the illumination 'I' of a firefly 'x' is selected as shown in Eq. (7.2).

$$I(x) \, \alpha f(x) \tag{7.2}$$

The light intensity or brightness of a firefly drops through the space after its basis and it remains immersed now in the media. So the bright intensity of a firefly becomes a function of both the distance and absorption coefficient of the media.

The bright concentration $I(r)$ of a firefly varying through both the distance r and absorption of the media is given by Eq. (7.3):

$$I = I0e - \gamma r2 \tag{7.3}$$

Here, 'I0' represents the light strength of a firefly by the source itself, and 'γ' represents the light immersion constant of the medium.

The attraction (β) among two fireflies, say 'i' and 'j', differs with the space between them. The firefly attractiveness is related to the illumination between two end-to-end fireflies; the attraction β between two fireflies is certain through Eq. (7.4).

$$\beta = \beta 0e - \gamma r\omega_s \tag{7.4}$$

where γ represents the space among two fireflies and $\beta 0$ represents the attraction between two fireflies at $r = 0$, i.e. two fireflies are situated by a similar point in the exploration space S. Now overall $\beta 0$ takes any value between 0 and 1.

The measure of a firefly 'i' involved toward a new additional smart (brighter) firefly 'j' is resolute through:

$$xi := xi + \beta 0e - \gamma r2ij = \beta(xj - xi) + \alpha \varepsilon i \tag{7.5}$$

Being anywhere as the randomization parameter, and remains a path of unplanned numbers pinched after Gaussian distribution or uniform distribution.

The fundamental operation in P2P-based networks is the lookup operation. Whenever a peer is looking for a data item, the query needs to be resolved efficiently and also successfully in less time. The main objective of any P2P-based wireless mesh network is to have low query response time, high packet delivery ratio and low network load, i.e., load balancing across wireless links. It is always required to find the best route for each query with a low query response time and a high packet delivery ratio. The objective of this work is to achieve the above-mentioned requirements in a P2P-centered wireless mesh network.

The proposed firefly algorithm for optimizing the performance of the chord protocol is given as follows:

FireFly Algorithm

FFA Meta-heuristic()
Begin
Initialize algorithm parameters
MaxGen: the maximum number of generations
γ: the light absorption coefficient
r: the particular distance from the light source
d: domain space
Define the objective function of f(x), where x = (x1, x2... xd)^T
Generate initial population of fireflies or x$_i$ (i = 1, 2...... n)
Determine the light intensity I$_i$ of x$_i$ via f(x$_i$)
While (t < MaxGen)
    For i = 1 to n (all n fireflies)
        For j = 1 to n (all n fireflies)
            If (I$_j$ > I$_i$)
                Move firefly i towards j by using equation 6.7
            End if
            Attractiveness varies with distance r via exp(-γr^2);
            Evaluate new solutions and update light intensity;
        End for j;
    End for i;
Rank the fireflies and find the current best;
End while;
Post process results and visualization;
End procedure

At first, the search difficulty jumps with an initial number of fireflies in the search space. Let the initial fireflies be considered as *xi*, *i*=1, 2, 3, …,*d*. Here, initial fireflies mean the initial possible set of nodes from the source to the destination. The brightness or intensity of each firefly is calculated by the objective function and is designated as Ii. That means the fitness of each route for the query is calculated. Among all possible routes, the route which gives the lowest end-to-end delay, highest package distribution part and lowest query response time is taken as the best route, i.e., the firefly that is the brightest among all the fireflies. The firefly through the smallest objective purpose worth takes top illumination since the optimization here remains to minimize the impartial purpose. All the fireflies will move toward the brightest firefly according to Eq. (7.7). New fireflies are generated and the brightness of new fireflies is calculated. This process is repeated until it finds the best route [5].

7.2.2 *Proposed Hybrid 'Particle Swarm Optimization-Firefly' (PSO-FF) Algorithm*

An optimization problem may generally consist of several local optima and global optimum. Though the FFA is an efficient optimization technique, sometimes it may fall into several local optimum points and it may not be able to come out of that. Due to this fact, it may not be able to search globally well in the search space of an optimization problem. The primary parameters of the FFA are the randomization parameter and absorption coefficient. The PSO algorithm is efficient in terms of convergence speed. This is mainly because it generates completely new random numbers r1 and r2 during each iteration step of the algorithm [6].

The proposed hybrid PSO-FF method combines the benefits offered by both PSO and FF (i.e., computational speed of PSO with FFA robustness) toward urging the worldwide searchability of algorithm in the search space. The algorithms PSO and FF are initiated through the first set of unsystematic solutions and only the finest explanation changes after one step toward one more centered arranged the principle of existence of fitting. This procedure continues until it reaches the maximum number of iterations or it meets any of the convergence criteria. An optimum solution is one which has the best maximum or minimum value of the objective function out of all sets of possible solutions [7]. The proposed PSO-FF algorithm performs local search with a light intensity operator of FFA and global search with a PSO operator.

Since the proposed method combined the advantages of both PSO and FF algorithms, it obtained better optimum results compared with results that were obtained with PSO and FF algorithms. Both PSO operator and FF operator are included in Eq. (7.8) to increase the search speed and also to enhance its global search capability. The proposed PSO-FF has almost similar iteration steps to those of the FFA except for that its position vector is updated as per Eq. (7.8).

The position vector xi of the PSO-FF algorithm is modified according to Eq. (7.8).

$$Vi(t+1) = wVi(t) + c1\text{rand}(.)(P * p\text{best}i(t) - Pi(t))$$

$$+ c2\text{rand}(.)(P * g\text{best}(t) - Pi(t)) \qquad (7.8)$$

The quasi encryption of the planned cross PSO-FF procedure is given as follows:

```
Begin
Generate the initial population of fireflies, xᵢ (i = 1, 2....... d)
Define the objective function f(x)
Determine the light intensity Iᵢ of xᵢ via f(xᵢ)
Initialize pbest and gbest
While (t < MaxGen)
For i = 1: d (d fireflies)
For j = 1: d
Distance between (pbest - xᵢ) and (gbest - xᵢ) are given by rₚₓ and rᵢ
If (I(j) > I(i))
FireFly i is moved towards FireFly j by using eqn 6.8
End if
Calculate new solutions and update light intensity value
Update pbest and gbest
End for j
End for i
End while
```

In the projected technique, the attraction between two fireflies is changed with the PSO operator. At this iteration step, every particle is attracted randomly toward the best in the complete population. The local search in the different regions of search space is done through the revised attractiveness step of the PSO-FF algorithm.

7.3 Performance and Simulation Results

For simulations, the amount of knobs is well-thought-out to be happening in the collection of 20–100. Table 7.1 gives the parameters used in the simulation environment. The imitations remain completed initially aimed at a stationary network than for the dynamic network where nodes are moving with different mobility, i.e., the performance of chord is examined by varying the network load and node mobility. The request interarrival time parameter indicates that the number of query messages are increased in the network with time to check the performance of the network with different network loads. The simulations are carried out in OPNET 14.5 simulator with back-end support of MATLAB for firefly and hybrid FFAs at the network layer.

From Tables 7.1 and 7.2 and Figure 7.1, it is observed that the proposed PSO-FF optimization increased the packet delivery ratio in the network by 10.75% and 9.18% when compared with Chord/FF and Chord/PSO optimization techniques, respectively.

From Table 7.3 and Figure 7.2, it is observed that the proposed PSO-FF optimization decreased the end-to-end delay in the network by 3.48% and 8.056% when compared with Chord/FF and Chord/PSO optimization techniques, respectively.

From Table 7.4 and Figure 7.3, it is observed that the proposed PSO-FF optimization decreased the network load in the network by 6.579% and 3.047% when compared with Chord/FF and Chord/PSO optimization techniques, respectively.

A. Dynamic network with different node mobility
From Table 7.5 and Figure 7.4, it is observed that the proposed PSO-FF optimization increased the packet delivery ratio in the

Table 7.1   Simulation Parameters for Hybrid Approach

SIMULATION PARAMETERS	VALUES
Deployment area	4 km × 4 km
Number of Fireflies	10
Random parameter α	0.5
Attractiveness $\beta 0$	0.2
Absorption coefficient γ	1
Maximum number of repetitions	100
MAC protocol	IEEE 802.11b
Query size	512 bytes
Node speed	20 kmph
Node mobility	Random way point
Simulation time	240 s

Table 7.2   Static Network by Varying Offered Load

REQUEST INTERARRIVAL TIME	CHORD/FF	CHORD/PSO	CHORD/PSO-FF
1	0.3	0.31	0.34
5	0.64	0.65	0.69
10	0.69	0.68	0.73
15	0.32	0.35	0.42
20	0.84	0.84	0.91

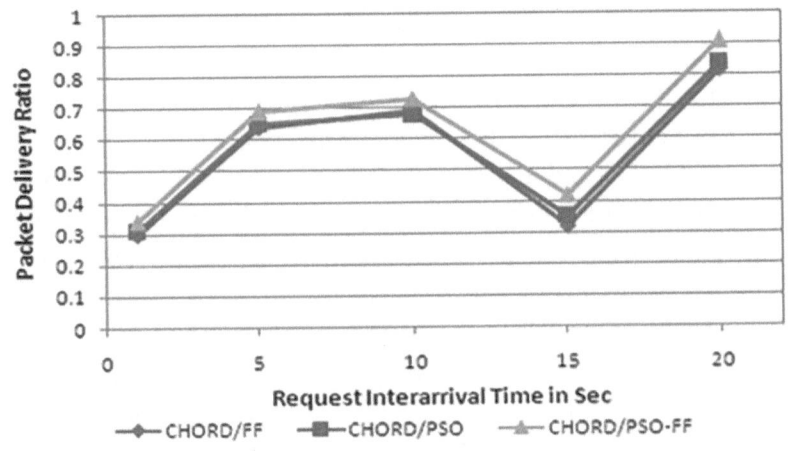

Figure 7.1   Packet delivery ratio (PDR).

Table 7.3   Request Interarrival Time

REQUEST INTERARRIVAL TIME	CHORD/FF	CHORD/PSO	CHORD/PSO-FF
1	0.31	0.33	0.29
5	0.38	0.4	0.37
10	0.4	0.43	0.38
15	0.44	0.45	0.43
20	0.48	0.5	0.47

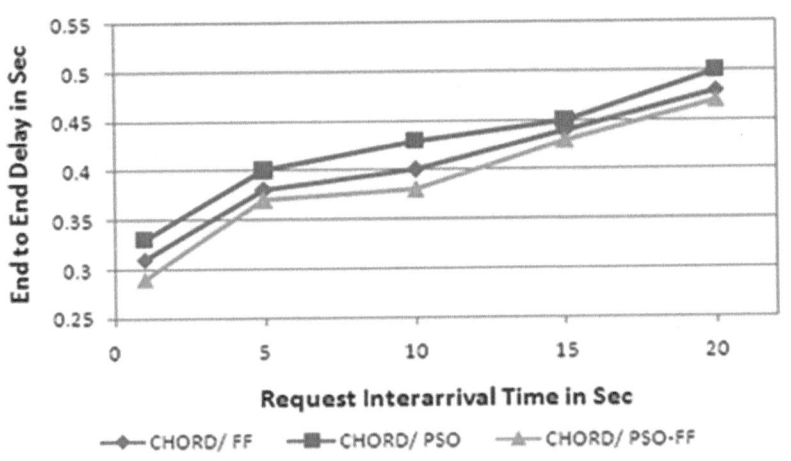

Figure 7.2   End-to-end delay in sec.

Table 7.4   Network Load in MB

REQUEST INTERARRIVAL TIME	CHORD/FF	CHORD/PSO	CHORD/PSO-FF
1	452	438	422
5	394	385	372
10	376	355	344
15	328	319	312
20	289	275	268

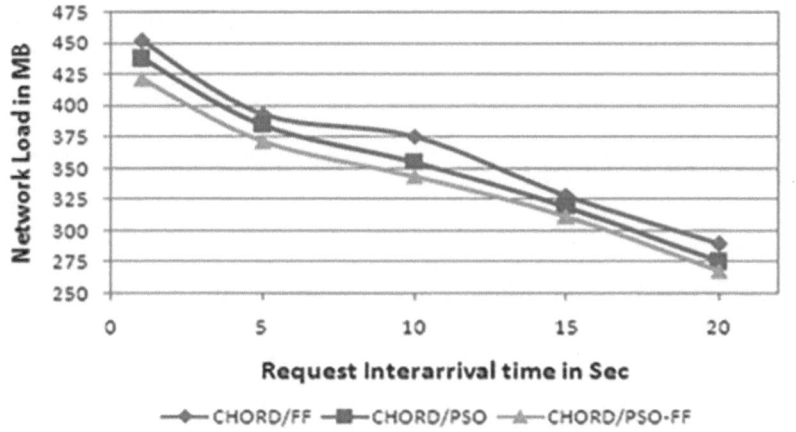

Figure 7.3   Network load in MB.

Table 7.5   Packet Delivery Ratio

NODE SPEED IN KMPH	CHORD/FF	CHORD/PSO	CHORD/PSO-FF
20	0.92	0.92	0.94
40	0.82	0.81	0.84
60	0.73	0.73	0.75
80	0.34	0.34	0.35
100	0.12	0.17	0.18

network by 4.43% and 3.03% when compared with Chord/FF and Chord/PSO optimization techniques, respectively.

From Table 7.6 and Figure 7.5, it is observed that the proposed PSO-FF optimization decreased the end-to-end delay in the network by 5.85% and 11.8% when compared with Chord/FF and Chord/PSO optimization techniques, respectively.

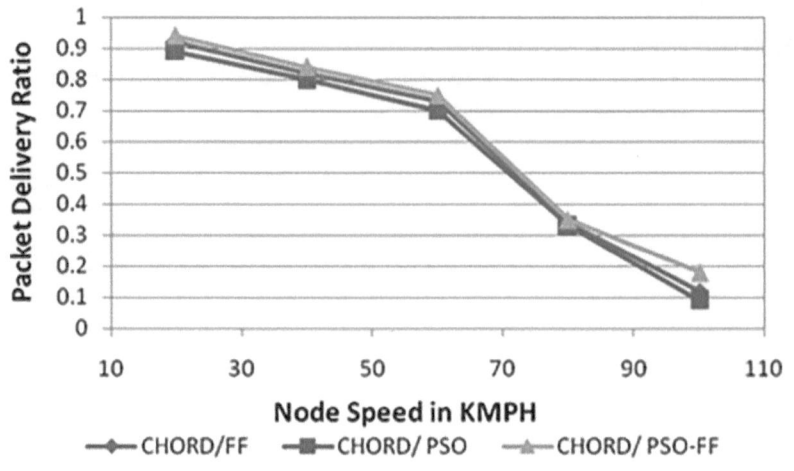

Figure 7.4 Packet delivery ratio (PDR).

Table 7.6 Node Speed in Kmph

NODE SPEED IN KMPH	CHORD/FF	CHORD/PSO	CHORD/PSO-FF
20	0.35	0.37	0.33
40	0.41	0.44	0.38
60	0.44	0.48	0.41
80	0.48	0.51	0.44
100	0.54	0.57	0.53

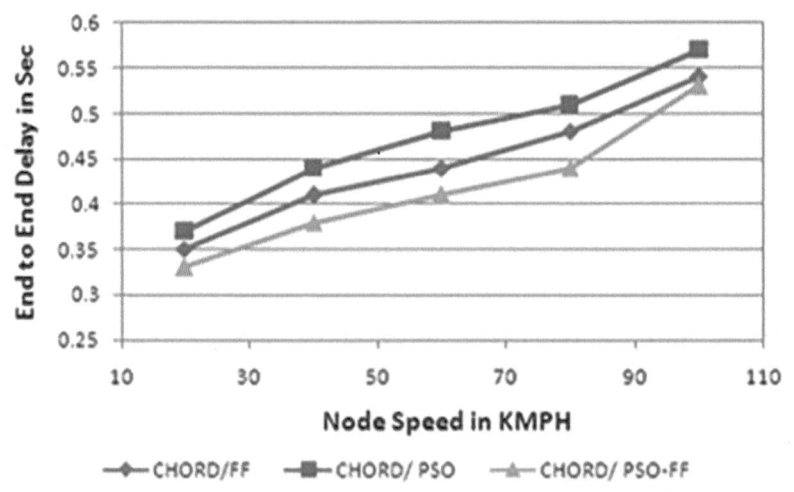

Figure 7.5 End-to-end delay in sec.

Table 7.7 Network Load in MB

NODE SPEED IN KMPH	CHORD/FF	CHORD/PSO	CHORD/PSO-FF
20	486.9	481.69	449.81
40	452.51	445.49	417.12
60	393.15	385.89	357.54
80	367.96	354.72	333.7
100	317.36	311.94	293.3

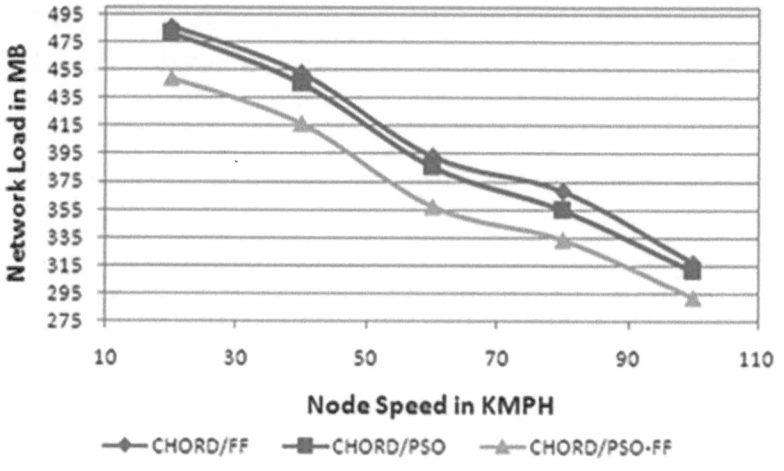

Figure 7.6 Network load in MB.

From Table 7.7 and Figure 7.6, it is observed that the proposed PSO-FF optimization decreases the network load in the network by 8.246% and 6.478% when compared with Chord/FF and Chord/PSO optimization techniques, respectively.

7.4 Conclusion

The chord protocol has been a widely deployed routing protocol for structured P2P overlay networks. The performance of these networks is influenced significantly by the routing protocol that they employ. It is shown in Chapter 4 that the performance of the chord protocol is degraded due to the dynamic nature of nodes when deployed over a P2P-centered wireless mesh network. In Chapter 5, the chord routing with PSO optimization is proposed which showed significant improvement in the performance of chord when compared with existing chord and location-aware chord. In this chapter, chord routing with FFA is proposed initially and then with the PSO-FF optimization technique. The FFA is an efficient optimization technique, but it may fall into several local optimum points and it may not be able to come out of that. Due to this fact, it may not be able to search globally well in the search space of an optimization problem. The proposed hybrid PSO-FF method combines the benefits offered by both PSO and FF algorithms (i.e., computational speed of PSO with FFA robustness) near surge the worldwide exploration capability of the algorithm in the search space. Since the proposed method has combined the advantages of both PSO and FF algorithms, the simulation results obtained are superior to those obtained with PSO and FF algorithms.

References

1. Maenpaa, J., & Camarillo, G. (2009, May). Study on maintenance operations in a chord-based Peer-to-Peer session initiation protocol overlay network. In *Parallel & Distributed Processing, 2009*. IPDPS 2009. IEEE International Symposium on (pp. 1–9). IEEE.
2. Yang, X.-S. (2013). Firefly algorithm: Recent advances and applications. *International Journal of Swarm Intelligence*, 1 (1).
3. Yang, X.-S. (2009). Firefly algorithms for Multimodal optimization, *International symposium on stochastic algorithms*. Springer Berlin Heidelberg.
4. Wang, G. G., Guo, L., Duan, H., & Wang, H. (2014). A new improved firefly algorithm for global numerical optimization. *Journal of Computational and Theoretical Nanoscience*, 11 (2), 477–485.
5. Murali Krishna P, P., Subramanyam, M. V., & Satyaprasad, K. (2016). A QoS improvement using hybrid swarm intelligence in Peer to Peer based Wireless Mesh Network. *Indian Journal of Science and Technology (IJST)*, 9 (36), 1–6, Sept. 2016.
6. Farahani, S. M., Abshouri, A. A., Nasiri, B., & Meybodi, M. R. (2012). Some hybrid models to improve firefly algorithm performance. *International Journal of Artificial Intelligence*, 8 (S12), 97–117.
7. Arunachalam, S., AgnesBhomila, T., & Babu, M. R. (2014). Hybrid particle swarm optimization algorithm and firefly algorithm based combined economic and emission dispatch including valve point effect. In *Swarm, Evolutionary, and Memetic Computing* (pp. 647–660). Springer International Publishing.

8

MESH-DHT Approach for Efficient Resource Sharing in P2P-Based Wireless Mesh Cognitive Networks

DAYADI LAKSHMAIAH, R. YADAGIRI RAO, AND A. SINDHUJA

Sri Indu Institute of Engineering and Technology

P. PRASANA MURALI KRISHANA

KITS College

S. POTHALAIAH

Vignana Bharathi Institute of Technology

Contents

8.1 Introduction

In the previous chapter, the organized P2P procedure chord is investigated above wireless mesh networks when nodes are stationary and mobile. The simulation results showed that the dynamic nature of nodes has a substantial effect on the QoS parameters. The P2P network practice is slightly small in a multi-hop environment such as wireless mesh systems. In multi-hop networks such as WMNs and MANETs, the dynamic nature of nodes and instability of a wireless medium will make a consistent loss of queries. The consistent chord ring is also difficult to maintain due to node mobility and node failures in a wireless network. The main difficulties in deploying P2P networks in a wireless mesh network are node mobility, bandwidth constraints, multi-hop forwarding, route stretching, request transmission and so on.

8.1.1 Methodology

In the basic chord protocol, the identifiers are given to each participating node without considering the physical locality of nodes. The successor of a node in the virtual ring may be far away in the actual physical network. The leaving node has to transfer all its keys to its next successor, which is actually far away in the network. This leads to unnecessary multi-hop forwarding of data, which also leads to congestion on the wireless link. The successor node transfers keys with IDs in the simulated loop. The awareness stays close through IDs to knobs that remain actually near in the real network. Thus nodes that are close in the actual physical network receive near identifiers. This assignment of nearby identifiers diminishes the above in the DHT statement, since knobs mostly exchange to those knobs that are physically close in the network, thus reducing the number of message transmissions. Due to this, a maximum of the messages remains wrapped among peers than their successor or predecessor in the virtual ring. This location awareness is very effective in reducing message overhead. A stable location-aware overlay network is taken, which enables fully distributed organization of information. The location-aware ID assignment is done through the use of GPS receivers of wireless nodes [1].

Proposed Particle Swarm Optimization (PSO) for QoS improvement in the previous work examined the feasibility of chord by deploying it over a wireless mesh network under a static environment, i.e. the end-users are stationary. It also considered the location of nodes to assign identifiers. The existing location-aware chord takes node locality into consideration to avoid unnecessary multi-hop forwarding of data. Several techniques have been proposed in the literature to enhance the routing efficiency of the chord to adapt it for different applications. Most of these techniques considered proximity neighbor selection, removal of redundant information in finger tables, topology-aware ID assignment and so on.

The performance of P2P overlay structures is mainly dependent on the routing protocol that they employ. We take the planned MESH-DHT approach that studies connection value, end-to-end interruption, request comeback period and package distribution proportion to war the QoS constraints. Since the optimization here is a multi-objective function, the atom group optimization technique is employed to enhance the limitations. None of the existing techniques has employed optimization algorithms on chord routing to improve its performance. The proposed work is different from the existing literature in that it employed optimization algorithms on chord routing. The next unit discusses the application of element group optimization to the proposed problem.

Dr Eberhart and Dr Kennedy industrialized the atom swarm optimization technique by taking the concepts of common behavior of bird gathering and fish training. The PSO algorithm jumps by an early random set of solutions in the exploration planetary and moves toward optima by updating the solutions. The awareness of the PSO remains emerged from the concepts of swarming behavior found in bird flocks and fish schooling. The PSO remains a resident-centered optimization instrument. The PSO can be applied effortlessly to answer numerous optimization difficulties in different fields [2].

The PSO algorithm is different from the Genetic Algorithm (GA) in that it does not contain the cross-over and mutation operators of the GA [3]. The PSO initially generates the population of particles randomly. The PSO algorithm performs a search by using a population of particles that are comparable to those used now in a GA. Every element consumes a place in the search space represented with a position path and difference between elements. These particles forward complete the search by updating their positions with a velocity represented with a velocity vector. Every particle checks and maintains its best position achieved so far after each iteration. The particle that has achieved the finest position among all particles is kept as the global best [4].

The succeeding suitability purpose remains active aimed at the PSO algorithm:

$$\min f(x) = d_{ijnQrt} \Bigg/ \left(\frac{\dfrac{L_{ij}}{eL_m} + 1}{e + 1} \right) \tag{8.1}$$

where d_{ijn} = regularized end-to-end interruption assumed through $d_{ijn} = \dfrac{d_{ij}}{d_m}$

d_{ij} = end-to-end interruption among the cause and the endpoint

d_m = extreme end-to-end interruption

L_{ij} = connection value among the basis and the endpoint

L_m = extreme connection excellence

Q_{rt} = stabilized request-response period

The Request Answer Period: It is the quantity of period occupied through a request created by a peer until the situation takes reply to ward that query. A regularized query answer period stands for the relation of query response time toward maximum query response time.

Packet Distribution Ratio: It is the relation between the number of packs delivered successfully and the number of packets directed.

End-to-End Interruption: This is a period that a data packet takes just before reaching its endpoint. This delay causes the interruption in the data packet transmission. The data packets that are situated effectively and transported near the destination are only considered.

The general PSO algorithm iteration steps are given in the current chart shown in Figure 8.1.

Quality: It is computed from the "Received Signal Strength Index" (RSSI) measured in dBm.

Effective lookup proportion: It is the relation between the number of requests effectively responded and the entire number of questions created in the network. This metric reflects the chord's ability to resolve a request consistently.

The major stages of the PSO algorithm are generating locations and speeding of the elements, location updates and speed information.

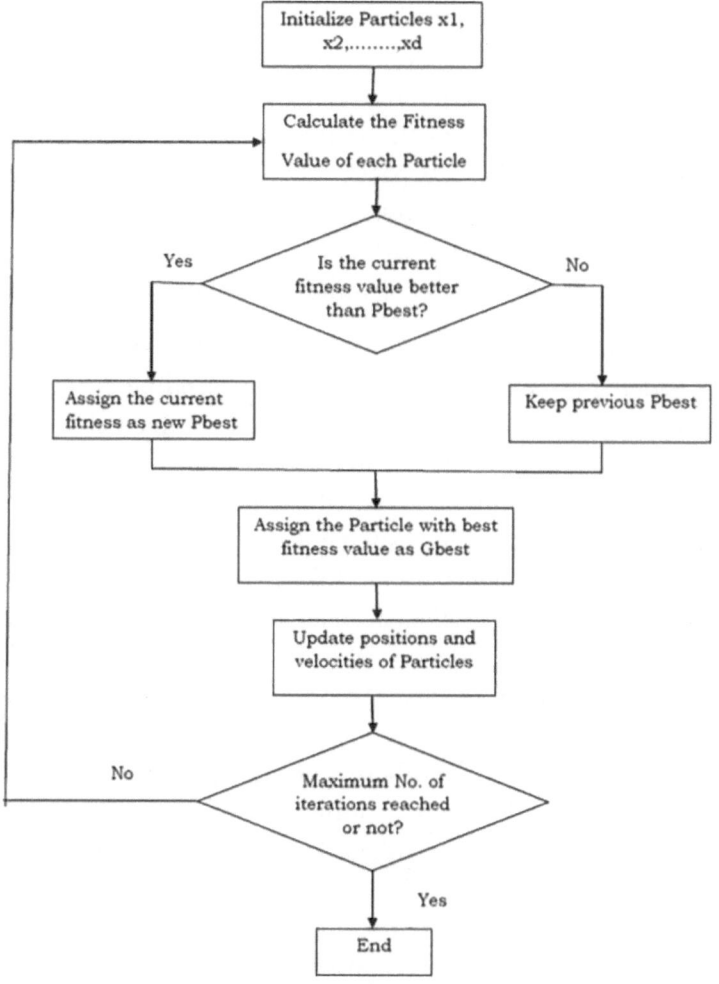

Figure 8.1 Flowchart of the PSO algorithm.

The element remains generally an opinion in the search space, which varies its position from one step to an additional step with the help of velocity updates. The particles move near the finest explanation that they take initially throughout every repetition [5].

The main requirements in a P2P-based wireless mesh network are to have low query response time, little network load, i.e. load balancing, great packet distribution proportion and successful query lookup. Efficiently locating a data item is one of the major objectives in P2P-based networks. Whenever a node issues a query for finding a data article, it has to be resolved in less time and also correctly. Unless a query is not resolved correctly, it becomes an unsuccessful query and such unsuccessful queries remain propagating in the network, which detains upsurge the circulation in the network and congestion on the relations. The best path from the source to the destination with low query response time and great packet delivery is always necessary for P2P-based wireless mesh networks.

The operation of the proposed PSO optimization procedure aimed at chord grounded wireless mesh network remains specified beneath [6]:

1. The difficult quest space twitches through first likely usual of nodes commencing basis node toward sending point node, named elements. The elements be presents sum edvia P={X1, X2 .... Xd}. Then all possible sets of routes from the basis toward the endpoint are taken which are called particles.

2. The procedure limitations such equally C1, C2, W, locations and speeds of elements are modified.

3. The suitability and designs deemed at all elements found happening relation excellence, end-to-end interval, packet distribution part then request-response time. Once this is done, the unit through the lowest suitability value situated named Gbest in addition to all others is called Pbest.

4. The location and speeds of units are efficient allowing the subsequent calculations: here C1 and C2 characterize acceleration coefficients, which take any value from 0 to 2; then r1 and r2 signify unsystematic figure snow among 0 and 1. Wherever 'i' differs after 1 toward 'd', 't' stands the repetition amount.

5. With efficient locations and speeds, new elements remain produced. The suitability of different units is designed.

6. On behalf of each unit, relate the present element's scalability worth with its previous Pbest. If this present rate remains improved than the previous Pbest value, then set this as Pbest; otherwise, keep it as it is. Similarly, compare global best of current particles with previous Gbest. If this value is improved, then proceeding before usual this is Gbest; otherwise, keep the previous Gbest.

7. The periods from 1 to 6 remain recurring till the extreme number of repetitions.

8.2 Proposed Work

In this chapter, Cognitive Radio (CR) is a piece of knowledge that purposes by refining spectrum efficiency in wireless communications through resourceful usage of the obtainable spectrum possessions. We plan and instrument a reasoning radio hardware stage created and arranged using Worldwide Software Radio Peripheral (USRP), which remains proficient in casing varied groups from 2.3 to 2.7 GHz and varying limitations. Power regulators situated are typically used for spectrum distribution CR schemes for making the most of the size of subordinate operators through intrusion power restraints to defend the main operators. The main requirements in a P2P-based wireless mesh network are to have low query response time, low network load, i.e. load balancing, high packet delivery ratio and successful query lookup. Efficiently locating a data item is one of the major objectives in P2P-based networks. Whenever a node issues a query for finding a data item, it has to be resolved in less time and also correctly. If a query is not resolved correctly, it becomes an unsuccessful query and such unsuccessful queries keep propagating in the network which in turn increases the traffic in the network and congestion on the links. The best path from source to destination with a low query response time and a high packet delivery is always necessary for P2P-based wireless mesh networks.

The working of the proposed PSO optimization algorithm for a chord-based wireless mesh network is given below:

1. The problem search space starts with the initial possible set of nodes from the source node to the destination node, called particles. The particles are given by $P=\{X_1, X_2 \ldots\ldots\ldots X_d\}$. Then all possible sets of routes from source to destination are taken which are called particles.
2. The algorithm parameters such as C_1, C_2, W, positions and velocities of particles are initialized.
3. The fitness is calculated for each particle based on link quality, end-to-end delay, packet delivery ratio and query response time. Once this is done, the particle with the lowest fitness value is called Gbest and all others are called Pbest.
4. The position and velocities of particles are updated according to the following equations:

$$v_i\left(t+1\right)= \omega v_i\left(t\right)+ c_1 r_1\left(P\text{best}_i - x_i\left(t\right)\right)+c_2 r_2\left(G\text{best}_i - x_i\left(t\right)\right) \quad (5.2)$$

$$x_i\left(t+1\right)= x_i\left(t\right)+ v_i\left(t+1\right) \qquad (5.3)$$

where c_1 and c_2 represent acceleration coefficients that take any value in between 0 and 2, and r_1 and r_2 represent random numbers in between 0 and 1. Here, 'i' varies from 1 to d and 't' is the iteration number.

5. With the updated positions and velocities, new particles are generated. The fitness of new particles is calculated.
6. For every particle, compare the present particle's fitness value with its previous P_{best}. If this current value is better than the previous P_{best} value, then set this as P_{best}, otherwise, keep it as it is. Similarly, compare global best of current particles with previous G_{best}. If this value is better than the previous, then set this as G_{best}, otherwise, keep the previous G_{best}.
7. The steps from 1 to 6 are repeated till the maximum number of iterations.

8.3 Performance and Simulation Results and Discussion

The number of nodules for simulations is now measured to be in the range of 25–100 and the flexibility of each node is 20 kmph. The simulations are done in OPNET 14.5 with the hindmost end support of MATLAB. The simulation parameters for the proposed PSO optimization algorithm for QoS improvement in P2P-based wireless mesh networks are presented in Table 8.1.

The recreation remains showed for

1. Chord
2. Existing position alert chord protocol
3. Projected PSO optimization

From Table 8.2 and Figure 8.2, it is observed that the planned PSO optimization completes the movement now in the network by 6.66% while being associated with the present place alert chord and by 25.33% after related through innovative chord correspondingly.

Table 8.1 Simulation Parameters for the MESH-DHT Approach

SIMULATION PARAMETER	VALUE
Deployment area	4 km × 4 km
C1	Any value between 0 and 2
C2	Any value between 0 and 2
W	0.9
No. of Particles	10
Maximum No. of Iterations	100
Query Size	512 bytes
Node Speed	20 kmph
MAC Protocol	802.11b
Node Mobility	Random way point

Table 8.2 Amount of Nodes

AMOUNT OF NODES	CHORD	POSITION AWARE CHORD	OFFERED PSO OPTIMIZATION
25	9	7	6
50	13	11	10
75	21	16	15
100	32	26	25

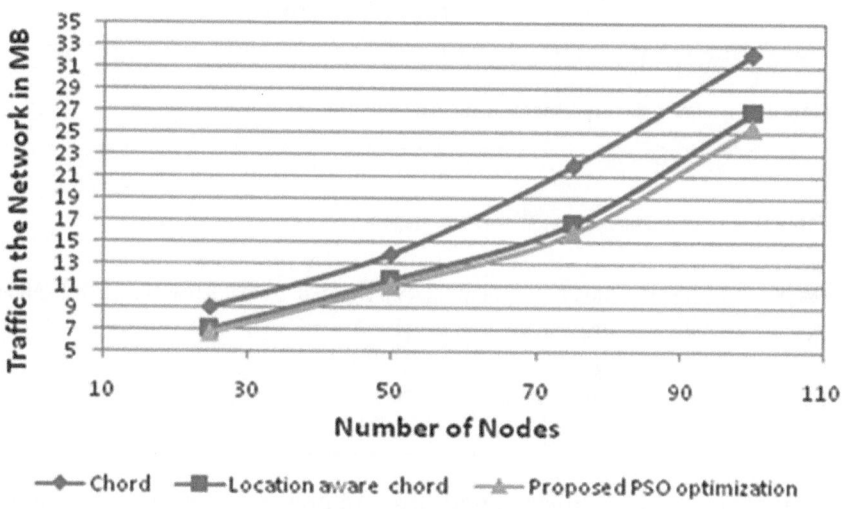

Figure 8.2 Movements in the network in MB.

From Table 8.3 and Figure 8.3, it is perceived that the offered PSO optimization condensed the normal request-response time in the

Table 8.3   Ordinary Request Reaction Time in Second

AMOUNT OF NODES	CHORD	POSITION AWARE CHORD	OFFERED PSO OPTIMIZATION
25	0.2624	0.2142	0.1976
50	0.322	0.2725	0.2567
75	0.395	0.3446	0.3198
100	0.5151	0.4295	0.4056

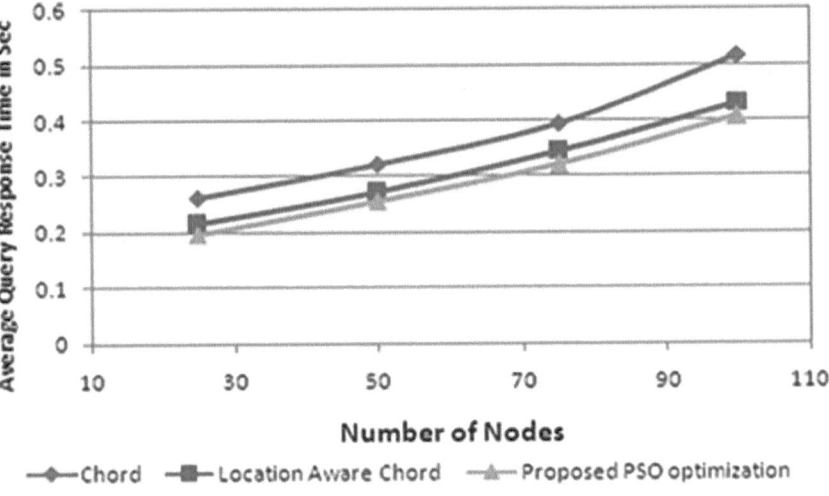

Figure 8.3   Average query response time.

Table 8.4   Successful Lookup Ratio in Percentage

AMOUNT OF NODES	CHORD	LOCATION AWARE CHORD	PROPOSED PSO OPTIMIZATION
25	88	91	93
50	83	88	89
75	76	79	80
100	69	73	76

network by 6.44% while equated through present position aware chord and by 21.06% when related through original chord correspondingly.

From Table 8.4 and Figure 8.4, it is observed that the proposed PSO optimization increased the successful lookup ratio in the network by 2.11% when compared with the present location-aware chord and by 6.96% when related through original chord respectively.

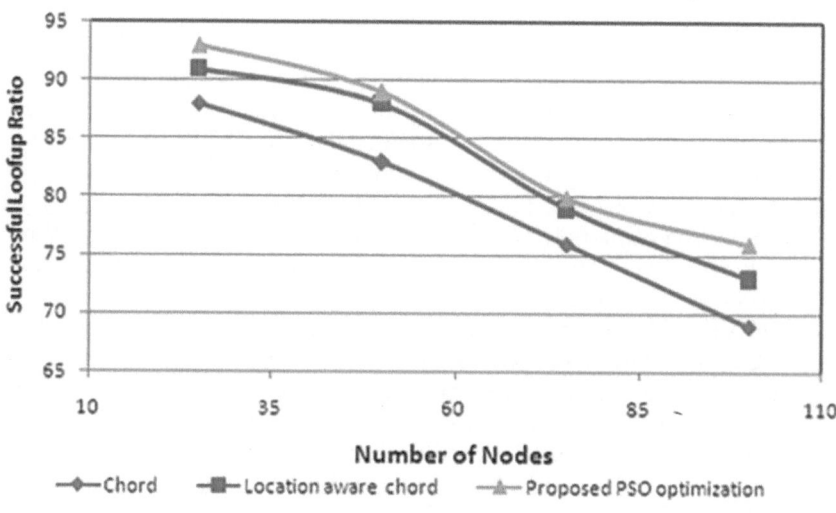

Figure 8.4 Successful lookup ratio in percentage.

8.4 Conclusion

The investigation of the chord protocol over a wireless mesh network demonstrated that the existing QoS limitations are degraded due to the dynamic nature of nodes as compared with static nodes. The locality of communication is essential in P2P-based networks to avoid the necessary forwarding of data and congestion on the wireless links. The existing location-aware chord takes node locality into consideration. The nearby nodes in the network thus take nearby identifiers. The MESH-DHT considers link quality, end-to-end delay, query response time and packet delivery ratio. PSO is proposed to improve the QoS parameters. The average query response time is reduced efficiently with the proposed PSO optimization when compared with the existing chord and location-aware chord. The successful query lookup ratio is also increased efficiently with proposed PSO optimization. Similarly, the traffic in the network is likewise condensed capably with the proposed PSO optimization. In addition to WLAN, the range detecting CR is added for measuring the gesture in the free space.

To provide the security in communications is one of the application of CR when the signals are transmitted in the white space.

References

1. Canali, C., Renda, M. E., & Santi, P., Evaluating load balancing in peer-to-peer resource sharing algorithms for wireless mesh networks, In Mobile Ad Hoc and Sensor Systems, 2008. MASS 2008. 5th IEEE International Conference on, sept 2008, pp. 603–609.
2. Hassan, R., Cohanim, B., De Weck, O., & Venter, G. (2005, April). A comparison of particle swarm optimization and the genetic algorithm. In Proceedings of the 1st AIAA multidisciplinary design optimization specialist conference (pp. 18–21).
3. Del Valle, Y., Venayagamoorthy, G. K., Mohagheghi, S., Hernandez, J. C., and Harley, R. G. (2008). Particle swarm optimization: Basic concepts, variants and applications in power systems. *Evolutionary Computation*, IEEE Transactions on, 12(2), 171–195.
4. Bratton, D., and Kennedy, J. (2007, April). Defining a standard for particle swarm optimization. In Swarm Intelligence Symposium, 2007. SIS 2007. IEEE (pp. 120–127). IEEE.
5. Aljober, M. N., and Thool, R. C. Multi-Objective Particle Swarm Optimization for Multicast Load Balancing in Wireless Mesh Networks.
6. Murali Krishna Prasanna, P., Subramanyam, M. V., and Satyaprasad, K. (2015). Mesh DHT approach for efficient resource sharing in P2P based Wireless Mesh Networks. *ARPN Journal of Engineering and Applied Sciences*, 10(21).
7. Maenpaa, J., & Camarillo, G. (2009, May). Study on maintenance operations in a chord-based Peer-to-Peer session initiation protocol overlay network. In Parallel & Distributed Processing, 2009. IPDPS 2009. IEEE International Symposium on (pp. 1–9). IEEE.
8. Yang, X.-S. (2013). Firefly algorithm: Recent advances and applications. *International Journal of Swarm Intelligence*, 1(1).

9

Design of the VLSI Technology for Cognitive Radio

DAYADI LAKSHMAIAH

Sri Indu Institute of Engineering and Technology

S. POTHALAIAH

Vignana Bharathi Institute of Technology

FARHA ANJUM

Siddhartha Institute of Engineering and Technology

A. SINDHUJA

Sri Indu Institute of Engineering and Technology

S. POTHALAIAH AND MOHAMMAD ILLIYAS

Shadan College of Engineering and Technology

Contents

DOI: 10.1201/9781003102625-9

9.1 Introduction

An electromagnetic two-way radio range shows a significant element inside the wireless message. By the side of the current, as of the arrival of the growth of an innovative wireless technology, the two-way radio range ruins individual use at length in addition to have expanded incomplete. The cognitive radio requires an advanced technology to perform very fast [1,2]. Administration performance gets flexible as the job of range sharing and their rule of still range allotment are as well added to the range shortage trouble. To undertake the difficulty of range contribution, a two-way radio range is approved with some rigid body for a variety of applications. After a definite bandwidth of range is owed to the major consumer, the major consumer resolves not to utilize the whole range, slightly a few regions are missing idle. These unused band locations are called range holes or white places. A subordinate client who does not take right toward using this range attempts to create usage of these white spaces which are idle through the main client not including invasive with the main client in the direction of weighing down the difficulties of range distribution, cognitive radio (CR) technology consumes remained planned by a technique of reason. The most important reason for CR ruins is to rush the process of range during classifying unused beside mode of a beneath-utilize band within jointly common ones after that rapidly changing environmental conditions [3]. Usually, four methods are normally throw-off designed for spectrum detecting. They are coordinated strain finding, energy detection, waveform centered detection and then cyclostationary characteristic discovery. Power recognition remains the maximum extensively recycled technique since this one is easy by the way it does not require some previous information approximately the effort indication. The power discovery method used here is aimed at condensed difficulty [4].

This chapter remains prearranged now the next method: segment II demonstrates an enhanced supple band detecting system toward dropping the difficulty of preceding structure and provides a report around the entire procedure intricate within this band sense technique. The VLSI construction of an RDDC and that one behind varying procedures remain clarified in segment III. The amount produced as of RDDC remains toward the FIR filter one through one as well as its process is clarified in segment IV. Segment V provides a thorough clarification around the procedure detecting the facts by power finding in addition to the VLSI application of liveliness control of the indication that is assumed. New consequences remain exposed within segment VI as well as after all tracked through assumptions.

9.2 Enhanced Flexible Range Detection System

It contains an RDDC, a humble FIR mesh and liveliness uncovering the founded detection scheme, which is shown in Figure 9.1. The elementary process of the whole preparation remains to principal cleverness the assumed contribution into sub-parts through depressed changing them by means of an RDDC [5]. Formerly all down-converted production remains clean by means of FIR filter. The filter production is lastly supposed to be the power finding once the power of the assumed information is found. The power is zero due to uncertainty and if a specific power charge is perceived next to the signal remains constant.

This arrangement has the following considerations:

1. RDDC remains specified by the mandatory move contribution, which signifies the digital method of the indication.
2. The filter remains secure by the pre-defined strainer coefficients.
3. The downwards transformed output remains before being assumed to FIR filter one next to one.
4. The found filter amount produced remains formerly assumed to vigor discovery aimed at the control of the power of the assumed facts.

Figure 9.1 Optimized supple band sense system.

9.3 VLSI Structural Design Used for RDDC

The straightforward reconfigurable building block aimed at cognitive two-way radio application is the digital downwards converter (DDC). The mandatory regularity group of attention exists, which is able to separate down the band with the aid of a digital down converter. This down converter is used to decrease the existing speed. Therefore, by using a DDC, the compulsory group exists, which is able to be chosen and its occurrence is able to be changed downwards to the baseband occurrence [6,7]. The production of the digital down converter resolves recall altogether, the evidence concerning our compulsory incidence group of attention nevertheless will change it downwards to baseband and hereafter the model occurrence is able to be importantly condensed.

Figure 9.2 illustrates the construction of an RDDC. RDDC comprises an oppose-AND gate-multiplexer preparation used for location/altering the situation occurrence, two collector units (Acc1 and Acc2) and a ROM aimed at stowing the sine otherwise cosine provisions.

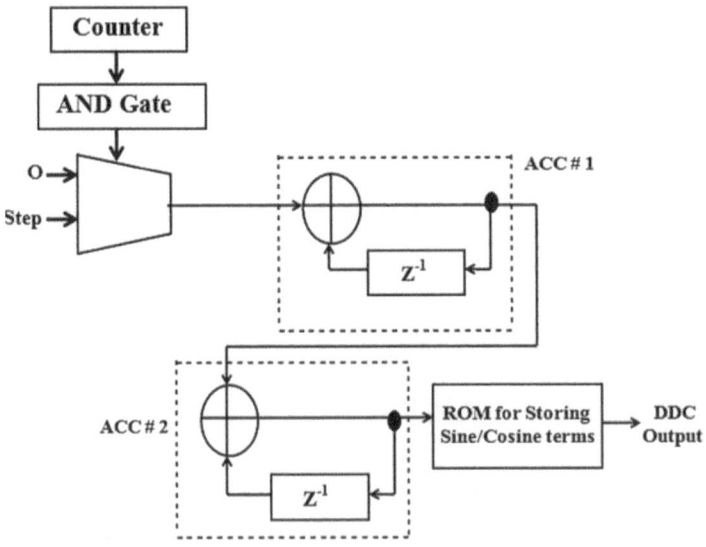

Figure 9.2   VLSI structural design of RDDC.

9.4 Performance of FIR Filter

The key function of FIR block is to down convert as well as changing the baseband signal. The filter identifies the signals and makes it as groups. So easy filter construction is used to improve the part and power constrain toward an important stage [8].

Digital filters usually use the same vital component of ordinary electronics. FIR filters remain individual of generally significant kinds of numerical filters used within digital signal processing. Some kind of occurrence reply can remain applied digitally through the FIR filter. It is normally useful through the aid of D flip flops, multipliers and adders shut near to gain the filter production. The essential block diagram of a customized FIR filter remains by way of giving away within the figure beneath. The D flip flop turns by way of an interruption module, which results in the procedure happening the preceding contribution signals. The h principles remain the constant standards [9,10]. These stay in use in the developmental procedure therefore with the aim of the amount produced by a specific period determination remain the amount of every deferred sample multiplied through their consistent coefficient standards. The method is complicated in choosing the distance of the filter as well as the constant standards remain recognized while the filter intends. The chief plan of the filter is to put these limits within such a method to find passband as well as stop band limits afterwards the process of the strain (Figure 9.3).

Depending on the extent, the strain reaction exists, which is able to remain enhanced. To facilitate incomes through with means of the filter construction by an extra number of blows, additional lightly tuned reply container is established [11,12]. By pre-defined duration in addition to coefficients, the activity of the filter is directly presumptuous.

A customized filter follows the subsequent steps:

1. Model the contribution sign found as of the RDDC.
2. Place in the latest example keen on the round barrier wherever the previous example is overwritten now by means of the new one.
3. Increase altogether the preceding sample by their suitable coefficient values. Currently, their amount has become the present amount produced.

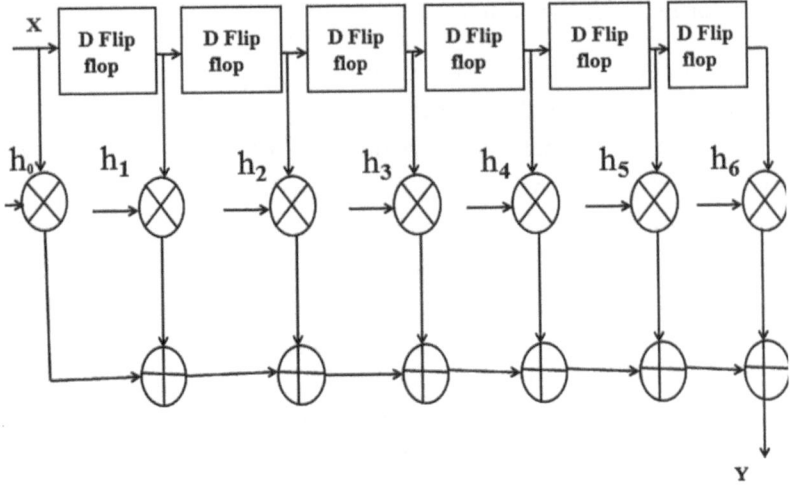

Figure 9.3 Performance of a FIR filter.

4. Recurrence the over procedure intended designed for all phases as well as the production of the filter remains attained next to the concluding stage.

Therefore, the over filter produces an impartial standard of its main current contribution sample. The input reason of the filter deceits inside the coefficient principles, which offer the definite production designed meant for an agreed prototype of contribution samples. Now the filter is applied for seven stages. It is difficult to filter the large amount of samples when the multiplexer is not used.

9.5 Energy Detection

Force recognition remains a modest indication recognition technique, which is mentioned as radiometry. An energy detector is appropriately meant for broadband range detection as soon as enough information around the PU signal cannot exist is collected through the CR. Force recognition remains the frequently suitable finding method as in most of the cases in sequence about the main operator is not recognized [13,14].

9.5.1 Performance of Energy Detection within an Instance Area

A multiplier, adder as well as monitor through a scale element are able to use applying the over Eq. (9.3) inside the digital method. On the way to becoming the square assessment of all illustrations, the inward signal samples leftovers because of contribution toward the multiplier. These square samples are gathered ensuing in the summary of square principles. An organized indication amount is continued within the collector, which produces the amount of square sample getting accumulated, i.e. the collector within contains an opposition consequently as a result of the calculated quantity of the samples being produced. To end with the procedure of averaging, the amount of the formed principles is approved over a scale unit. Therefore, the amount produced of a scale unit is the force of the contribution digital signal. The force sensor building obstruct illustration is given in Figure 9.4.

The primary VLSI construction of the power detector is shown in Figure 9.5. To become the square principles, the effort models are

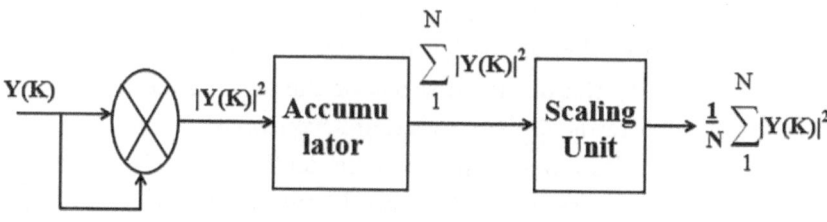

Figure 9.4   Block illustration of power detection.

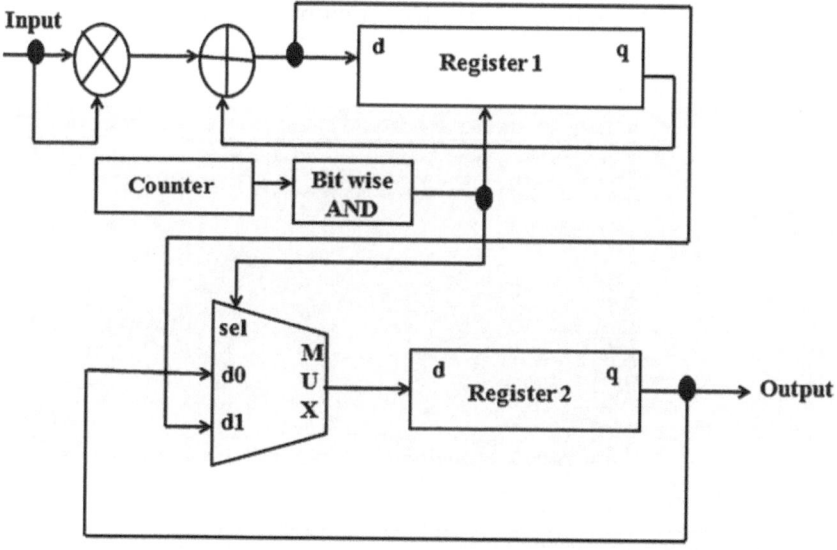

Figure 9.5   VLSI construction used for an energy detector.

specified toward a multiplier. In addition to register circuit, adder is used as a collector. The collector retains on top of adding up its inputs until the record obtains a manage signal behind the oppose. The oppose in addition to bitwise AND gate output sense elevated once the amount of the sample ranges the chosen worth. This organized sign is assumed toward the change of the collector, i.e. the collector starts as of zero once the wanted count of models is accumulated. The energy detector detects the samples energy at any time in the spectrum and gives the decision based on energy present in the sample.

9.6 Experimental Result

The potential improved supple group detects preparation as applied in the XILINX ISE emulator via a way of Verilog coding. The imitation meant in totaling the blocks is completed discretely besides the entire system is together applied. An action contribution of 128 bit is assumed to signify the occupied series of the sign inside the digital procedure. The filter is applied aimed at seven-phase as well as eight contribution samples remain known since dealing with the ROM in RDDC. These addresses choose the production of the DDC. At last, the force detector bounces the force rate of the indication, which controls the company otherwise lack of signal.

The DDC process is done as shown in Figure 9.6 and 128 spot move contribution remains separated in eight dissimilar 16 bit down

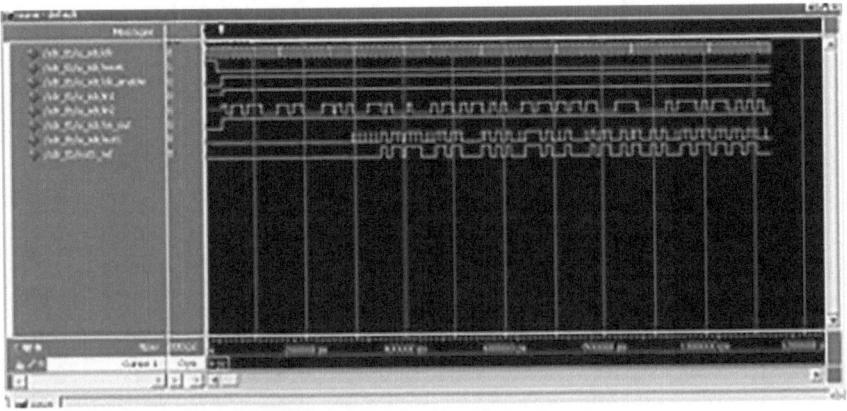

Figure 9.6   Model result of RDDC.

transformed outputs. Therefore, the address known decides the production of DDC. The design of the RDDC is to serve towards the filter one next to one. At present, the filter is intended for seven phases as well as the production is found from end to end to be around altogether in the preceding stage output. The last amount produced by the filter is shown in Figure 9.7. Therefore, the average amount of construction produced is 32-bit as well as is a well-organized filter amount produced.

The filter amount produced is assumed to be the force detected. It squares the input after that increasing is ended by means of a reproduce with mount up training. At last, since this preparation, the liveliness worth of the indication is found. The power finding block analyses the force of the supposed effort as shown in Figure 9.8. But the energy worth is 0; formerly the signal situated is said to be absent. Uncertainly, some energy value is found aimed at the sign; formerly it is supposed with the intention of the sign that is present.

The overhead distinct block remains applied together. The concluding amount produced of the improved flexible spectrum detecting system is shown in Figure 9.7. In Figure 9.7, the energy charge of the facts is zero meant designed intended for a convinced period as well as certain power charge is found on an additional instance. Zero rates specify the lack of the signal besides energy worth specifying the occurrence of the indicator. Consequently, after a digital step input

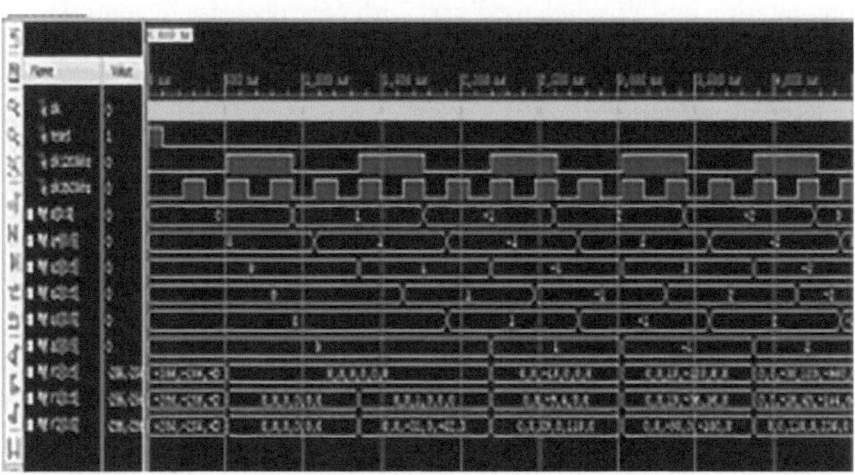

Figure 9.7   Model results of a FIR filter.

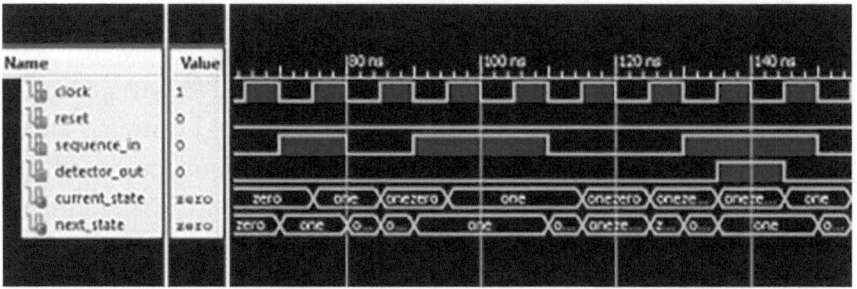

Figure 9.8 Model result of a power detector.

is assumed toward the sensing system, depending on the deal with designated, it computes the force of the signal and then expresses the occurrence of the signal.

The test of the outcome, influence and information pathway wait in the totaling region is planned to be intended for a supple sense system [3] and optimized supple sense scheme and assessment stay complete together the schemes. The assessments are shown in Table 9.1.

The power study is experimental in the XpowerAnalyzer by means of a Xilinx device. The power extreme through the optimized supple sense system is condensed while associated with the supple sensing system. The range employed with LUTs, IOBs and the amount of slice use remains practical as of the plan summary. The facts path delay remains distinguished since the synthesis report. The overhead outcomes show that the enhanced flexible sensing system deals improved presentation by 13% condensed part with 66% decrease in control.

Table 9.1 Execution Results

OPTIMIZED FLEXIBLE SENSE SYSTEM	SUPPLE SENSING SYSTEM	PARAMETERS
0.034	0.100	Power (W)
522	558	Amount of slice registers
1,271	1,496	Amount of portion LUTs
328	425	Amount of completely used LUT-FF pair
167	170	Amount of bond IOBs
5	5	Amount BUFG/BUFGCTRLs
7	12	12 quantity of DSP48E1s
2,300	2,666	Entire region
25.075	25.136	Delay (ns)
171,332	172,228	Memory procedure (KB)

9.7 Conclusion

It grants an innovative enhanced supply detecting system on the way to decrease the part as well as difficulty by means of a humble filter construction by D flip flops, multipliers and adders. The extent, control and interruption restraints are experiential by means of an XILINX ISE simulant. The enhanced flexible detecting arrangement proposals improved presentation by a condensed part when associated with the flexible range detecting scheme thus creating an extra humble and well-organized construction. In the future, the effort can remain long-lasting by applying this process by means of a helpful band sense. This sensor system can likewise remain realized by means of the additional novel finding methods similar to Eigen value-based discovery for finding an improved performance by near to the ground signal toward noise ratio (SNR).

References

1. J. Mitola and G. Q. Maguire, "Cognitive radio: Making software radios more personal," *IEEE Personal Communications.*, vol. 6, no. 4, pp. 13–18, Aug. 1999.
2. T. Yucek and H. Arslan, "A survey of spectrum sensing algorithms for cognitive radio applications," *IEEE Communications Surveys & Tutorials.*, vol. 11, no. 1, pp. 116–130, First Quarter, 2009.
3. R. Mahesh and A. P. Vinod, "A low-complex flexible spectrum sensing scheme for mobile cognitive radio terminals," *IEEE Transactions on Circuits and Systems-ii*, vol. 58, no. 6, pp. 371–375, Jun. 2011.
4. Y. Zhao, S. Li, N. Zhao and Z. Wu, "A novel energy detection algorithm for spectrum sensing in cognitive radio", *Information Technology Journal*, pp. 1659–1664, Asian network for scientific information, 2010.
5. M. López-Benítez and F. Casadevall, "Improved energy detection spectrum sensing for cognitive radio," *IET Communications.*
6. Y. Zeng and Y. C. Liang, "Eigenvalue-based spectrum sensing algorithms for cognitive radio," *IEEE Transactions on Communications*, vol. 57, no. 6, pp. 1784–1793, Jun. 2009.
7. B. Farhang-Boroujeny, "Filter bank spectrum sensing for cognitive radios," *IEEE Transactions on Signal Processing*, vol. 56, no. 5, pp. 1801–1811, May 2008.
8. R. Mehra and S.S. Pattnaik, "Reconfigurable design of GSM digital down converter for enhanced resource utilization," *International Journal of Computer Applications*, Volume 57, No.11, pp. 41–47, November 2013.

9. "Notice of proposed rulemaking and order: Facilitating opportunities for flexible, efficient, and reliable spectrum use employing cognitive radio technologies," FCC, Washington, DCET Docket No. 03–108, Feb. 2005.

10. J. Proakis, *Digital Communications*, 3rd edition. New York: McGraw-Hill, 1995.

11. H. Arslan and T. Yucek, "Spectrum sensing for cognitive radio applications," in *Cognitive Radio, Software Defined Radio, and Adaptive Wireless Systems*, H. Arslan, Ed. New York: Springer-Verlag, 2007.

12. Vipin. VHDL coding tips and tricks. [Online]. http://vhdlguru.blogspot.in/p/about-me.html

13. S. Haykin, "Cognitive radio: Brain-empowered wireless communications," *IEEE Journal of Selected Areas in Communication*, vol. 23, no. 2, pp. 201–220, Feb. 2005

14. J. Dalai, "VLSI Implementation of Energy Detection Algorithm for WLAN and WiMAX Applications," M.tech. thesis, 2013.

Index